SpringerBriefs in Molecular Science

For further volumes:
http://www.springer.com/series/8898

Francois Mathey

Transition Metal
Organometallic Chemistry

 Springer

Francois Mathey
Chemistry and Biological Chemistry
Nanyang Technological University
Singapore
Singapore

ISSN 2191-5407   ISSN 2191-5415   (electronic)
ISBN 978-981-4451-08-6   ISBN 978-981-4451-09-3   (eBook)
DOI 10.1007/978-981-4451-09-3
Springer Singapore Heidelberg New York Dordrecht London

Library of Congress Control Number: 2012955236

Printed on acid-free paper

Springer is part of Springer Science+Business Media (www.springer.com)

Preface

Today, chemistry textbooks tend to become bigger and bigger, following the development of the field. This trend has two consequences: these books become more and more useful for researchers and, at the same time, more and more frightening for students. After having taught transition metal chemistry for more than 20 years in France, California, and Singapore, I am convinced that there is room for a concise textbook focusing on the main products, reactions, and key concepts of the field. This philosophy means that this book necessarily will not be comprehensive and will treat only the core of the subject. In practice, the text is based on the course given to the students of NTU. Brevity does not mean superficiality. The level of this book is not elementary. Whenever possible, it blends theoretical explanations and experimental description. The student using this book should know basic organic chemistry and molecular orbital theory. In spite of its conciseness, it is hoped that this book will help students to quickly grasp the essence of the current developments in the field. Finally, I would like to acknowledge the help of Dr. Matthew P. Duffy who read the initial manuscript and suggested some improvements and all those who worked on the proofs. I dedicate this book to my wife Dominique who faithfully supported me during a long and sometimes difficult career.

Singapore, August 2012   Francois Mathey

Contents

Abbreviations

acac	Acetylacetonate	$[CH_3C(O)CHC(O)CH_3]^-$
bipy	2,2′-bipyridine	
Bu	*n*-butyl	
cod	1,5-cyclooctadiene	
Cp	Cyclopentadienyl	C_5H_5
Cp*	Pentamethyl-Cp	C_5Me_5
Cy	Cyclohexyl	
δ	Chemical shift (NMR)	
diphos		$Ph_2P-CH_2CH_2PPh_2$
DMF	Dimethylformamide	Me_2CHO
DMSO	Dimethylsulfoxide	Me_2SO
Et	Ethyl	C_2H_5
HMPT	Hexamethylphosphoro-triamide	$(Me_2N)_3P{=}O$
HOMO	Highest occupied molecular orbital	
L	Neutral 2-electron ligand	
LUMO	Lowest unoccupied molecular orbital	
Me	Methyl	CH_3
MO	Molecular orbital	
NHC	*N*-heterocyclic carbene	
ν	Frequency (IR)	
OAc	Acetate	$CH_3C(O)O-$
OS	Oxidation state	
Ph	Phenyl	C_6H_5
py	Pyridine	C_5H_5N
TBP	Trigonal bipyramid	
THF	Tetrahydrofuran	C_4H_8O
TMEDA	Tetramethylethylenediamine	$Me_2N-CH_2CH_2-NMe_2$
TMS	Tetramethylsilane	Me_4Si
T.O.F.	Turnover frequency (catalysis)	
T.O.N.	Turnover number (catalysis)	
X	Anionic 1-electron ligand	

Chapter 1
General Topics

Abstract This introductory chapter starts by a brief history of the subject from the discovery by Zeise of a platinum-ethylene complex in 1827 to the last Nobel prizes awarded to Heck, Negishi, and Suzuki in 2010 for their work on palladium-catalyzed carbon–carbon coupling reactions. Then, the electronic characteristics of the transition metals are presented (number of d electrons, electronegativities), together with the shapes of the atomic d orbitals. The various types of ligands are introduced with their coordination modes, terminal, bridging, mono- and poly-hapto. The special cases of CO, NO are discussed. The molecular orbitals of ML_6, ML_5, ML_4, ML_3, and ML_2 complexes are qualitatively studied. In each case, the structure of the d block is deduced from that of ML_6 using simple geometrical arguments. The main types of reactions of transition metal complexes are defined, including substitution, oxidative addition, reductive elimination, oxidative coupling, reductive decoupling, 1, 1 and 1, 2 migratory insertions, nucleophilic and electrophilic attacks on coordinated ligands. For each type, the main mechanisms are discussed with their consequences for the electronic structures of the complexes. All this introductory material can serve to decipher the modern literature on transition metal chemistry together with its applications in catalysis and synthetic organic chemistry.

Keywords Transition metals • d orbitals • 18-electron rule • Ligand field theory • Reaction mechanisms

1.1 Some Historical Facts

It is not an exaggeration to consider 1828 as the birthday of modern chemistry. It was in this year that Wöhler, a German chemist, accidentally discovered that heating ammonium carbonate, a common inorganic substance, transformed it into urea, a typical organic compound. He thus, established the first unambiguous link between inorganic and organic chemistry and killed the vital force theory that was supposed to control organic chemistry. This founding event was followed by a fast and continuous development of organic chemistry.

F. Mathey, *Transition Metal Organometallic Chemistry*,
SpringerBriefs in Molecular Science, DOI: 10.1007/978-981-4451-09-3_1,
© The Author(s) 2013

Almost at the same time, Zeise, a Danish chemist working at the university of Copenhagen, discovered the so-called Zeise's salt K[PtCl$_3$(C$_2$H$_4$)], which can be obtained by bubbling ethylene into a water solution of K$_2$PtCl$_4$. This compound contained the first three-center η^2 bond between ethylene and platinum but this structure was not definitively established before 1969 by X-ray crystal structure analysis. At the time of its discovery, this compound remained a curiosity and did not induce any significant development of transition metal chemistry.

Much later in 1890, Mond, a German chemist working in England, discovered the reaction of carbon monoxide with nickel which leads to nickel tetracarbonyl [Ni(CO)$_4$] and patented the process for the purification of nickel based on the conversion of crude nickel into pure [Ni(CO)$_4$]. This became a widely used industrial process, but it did not induce a notable interest from the academic chemists because [Ni(CO)$_4$] is a low-boiling and highly toxic liquid.

In 1893, Werner, working at the University of Zurich, proposed the correct ionic structure for the adduct between ammonia and cobalt trichloride [Co(NH$_3$)$_6$]Cl$_3$ with a hexacoordinate central metal and laid the foundations of modern coordination chemistry. He was awarded the Nobel prize in 1913 for this work.

In 1925, the Fischer–Tropsch process converting a mixture of CO + H$_2$ into hydrocarbons was introduced. It uses heterogeneous cobalt or iron catalysts and can provide a gasoline substitute made from coal. It could become a major process when oil resources are exhausted.

In 1938, Roelen in Germany discovered the cobalt-catalyzed hydroformylation of olefins (or "oxo" process) which converts alkenes into aldehydes by formal addition of H...CHO onto the C=C double bond. This remains today one of the major processes of the chemical industry. More than 6 million tons of "oxo" products are synthesized each year.

In 1951, Pauson and Kealy accidentally discovered ferrocene [Fe(C$_5$H$_5$)$_2$] as a stable orange solid but were unable to establish its correct structure. Its genuine structure in which iron is sandwiched between the two cyclopentadienyls with ten identical Fe–C bonds was independently established one year later by Wilkinson and Fischer who were awarded the Nobel prize in 1973 for their work on sandwich compounds.

The titanium-catalyzed polymerization of olefins (mainly ethylene and propene) was introduced in 1955 by Ziegler and Natta and has revolutionized our everyday lives. Around 100 million tons of these polymers are produced each year. Ziegler and Natta were awarded the Nobel prize in 1963.

Then, an almost continuous flow of discovery took place. Among them, the first carbene complexes by Fischer in 1964, the metathesis of olefins around 1964, the so-called Wilkinson catalyst for the hydrogenation of olefins in 1965, and so on. This extraordinary dynamism of transition metal organometallic chemistry was rewarded by several Nobel prizes: in 2001, Knowles, Noyori and Sharpless for asymmetric catalysis, in 2005, Chauvin, Grubbs and Schrock for the metathesis of olefins and in 2010, Heck, Negishi and Suzuki for the palladium-catalyzed cross-coupling reactions in organic synthesis.

Zeise's salt $K[PtCl_3(C_2H_4)]$

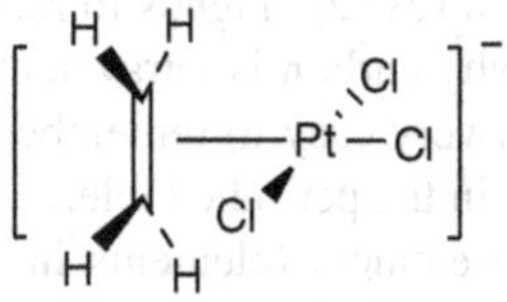

$$\left[\begin{array}{c} \overset{H \quad H}{\underset{H \quad H}{\bigvee}} \!\!\!-\!\! \overset{Cl}{\underset{Cl}{Pt}}\!\!-\!\!Cl \end{array} \right]^{-}$$

Mond process

$$\text{Ni (impure)} + 4\,\text{CO} \xrightarrow{50\,^\circ\text{C}} [\text{Ni(CO)}_4] \xrightarrow[230\,^\circ\text{C}]{\Delta} \text{Ni (pure)} + 4\,\text{CO}$$

Fischer–Tropsch process

$$\text{CO/H}_2 \xrightarrow{\text{cat. (Fe, Co)}} \text{hydrocarbons}$$

Oxo process

$$\text{RCH=CH}_2 + \text{CO} + \text{H}_2 \xrightarrow{\text{Co cat.}} \text{RCH}_2\text{CH}_2\,\text{CHO}$$

Ferrocene

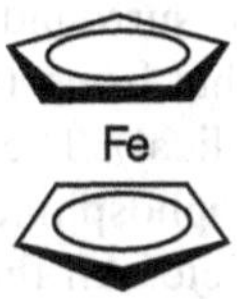

Ziegler-Natta process

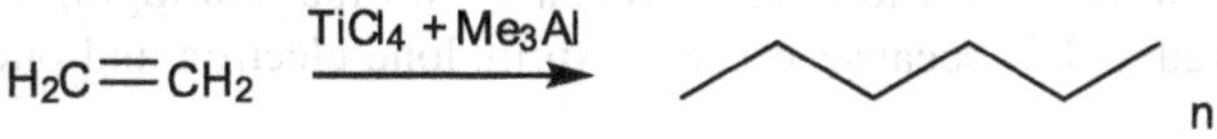

$$\text{H}_2\text{C=CH}_2 \xrightarrow{\text{TiCl}_4 + \text{Me}_3\text{Al}} \text{/\/\/\/}_n$$

Olefin metathesis

$$2\,\text{RCH=CH}_2 \rightleftharpoons \text{RCH=CHR} + \text{CH}_2\text{=CH}_2$$

1.2 Basic Data

A main group element such as carbon, nitrogen, etc., reacts through the electrons of its outside shell (n) and has typically the electronic configuration $ns^2\,np^x$ ($0 \leq x \leq 6$). Through its reactions, it tends to complete the np subshell at six electrons to reach the highly stable configuration of the next noble gas $ns^2\,np^6$. For example, carbon ($2s^2\, 2p^2$) tends to reach the configuration of neon ($2s^2\, 2p^6$). This is the reason why carbon is tetravalent. This trend is at the origin of the so-called octet rule. It is very easy to remember the value of n since it is identical with the period number in the periodic table.

A transition element (elements in the rectangle in the periodic table) such as titanium, iron, etc., is characterized by the fact that its $(n-1)d$ subshell has almost the same level of energy as its outside shell (n). Thus, this element reacts through the electrons of its outside shell (n) and the electrons of its $(n-1)d$ subshell. It has typically the electronic configuration $(n-1)d^x\, ns^2\, np^0$ ($0 \le x \le 10$). Through its reactions, it tends to complete the $(n-1)d$ and np subshells at ten and six electrons, respectively, to reach the highly stable configuration of the next noble gas $(n-1)d^{10}\, ns^2\, np^6$. For example, vanadium ($3d^3\, 4s^2\, 4p^0$) tends to reach the configuration of krypton ($3d^{10}\, 4s^2\, 4p^6$). This trend is at the origin of the so-called 18-electron rule which governs the chemistry of the transition metals. It must be noticed that vanadium can also lose its $3d^3$ and $4s^2$ electrons to reach the configuration of argon. This is the reason why vanadium is pentavalent. Organometallic chemists do not make a distinction between the s, p and d electrons since all of them participate to the chemistry of the transition metal. Conventionally, all of these are considered as d electrons. Hence, vanadium is viewed as a d^5 metal. It is very easy to remember this d count since it is identical to the group number in the periodic table.

When considering a complex, the first thing to do is to count the electrons around the metal. In order to do that, we need to know how many electrons are shared with the metal by the ligands surrounding it. Conventionally, two broad classes are distinguished, ligands that bring two electrons (lone pair, π–bond) and those bringing one electron (radical). The first class is called L, the second X. Examples of L are CO, amines, phosphines, singlet carbenes; examples of X are chlorine, alkyl, aryl, alkoxy, etc. All the other ligands are considered as a superimposition of L and X ligands. For example, allyls are LX, dienes are L_2, cyclopentadienyls are L_2X and arenes are L_3 when bonded by all of their carbon atoms to the transition metal. When a chlorine is bridging two metals, it is considered as LX because it uses both its lone electron and one lone pair

Simplified Periodic Table of the Elements

	1	2	3	4	5	6	7	8	9	10	11	12	13	14	15	16	17	18
1	H																	He
2	Li												B	C	N	O	F	Ne
3	Na												Al	Si	P	S	Cl	Ar
4	K	Ca	Sc	Ti	V	Cr	Mn	Fe	Co	Ni	Cu	Zn	Ga	Ge	As	Se	Br	Kr
5	Rb	Sr	Y	Zr	Nb	Mo	Tc	Ru	Rh	Pd	Ag	Cd	In	Sn	Sb	Te	I	Xe
6	Cs	Ba	La	Hf	Ta	W	Re	Os	Ir	Pt	Au	Hg	Tl	Pb	Bi	Po	At	Rn
7	Fr	Ra	Ac															

La and lanthanides; Ac and actinides

Table: Electron count for common ligands

Ligand	Covalent	Ionic	Equivalence	OS of M
H, R, Ar, F, Cl, Br, I, CN, RO, RS, R_2N, R_2P	1	2 (X^-)	X	+1
CO, OR_2, SR_2, NR_3, PR_3	2	2	L	0
Bent NO	1	2 (NO^-)	X	+1
Linear NO	3	2 (NO^+)		−1
μ_2–X (X=Cl, Br,RS, R_2N, R_2P)	3	4	LX	+1
=O, =S	2	4(O^{2-}, S^{2-})	X_2	+2
$=CR_2$ (electrophilic)	2	2	L	0
$=CR_2$ (nucleophilic)	2	4(CR_2^{2-})	X_2	+2
η^2-alkene	2	2	L	0
η^3-allyl	3	4(allyl$^-$)	LX	+1
η^5-cyclopentadienyl	5	6(Cp$^-$)	L_2X	+1
η^6-arene	6	6	L_3	0

for the coordination. The number of X ligands around a metal gives its oxidation state (OS). M–L and M–M bonds are not taken into account. Do not forget the positive or negative charges borne by the metal in this count. For example, ferrocene is equivalent to FeL_4X_2, hence, OS = +2; $[Fe(CN)_6]^{3-}$, OS = +3. Finally, when a ligand is bonded to a single metal by n atoms it is called n-hapto (η^n). For example, in ferrocene, the cyclopentadienyls are penta-hapto (η^5). When a ligand is bridging n metals by a single atom, it is labeled μ_n.

Two important ligands need special comments. Carbon monoxide (CO) is isoelectronic with dinitrogen (N_2). It contains a triple bond between C and O and two axial lone pairs on C and O. Its representation is $^-C \equiv O^+$. It almost always coordinates to a metal through C. Nitrogen monoxide (NO) is a radical. When using its lone electron, it gives a bent M–N=O complex (OS = +1). When using the three electrons on nitrogen, it gives a linear M=N=O complex in which NO is considered as equivalent to NO^+. Hence, in this complex the OS of the metal is −1. These data are summarized in the table.

1.3 Electronic Structures

Broadly speaking, the chemistry of transition metals is the chemistry of d electrons. The shapes of the five d orbitals is given below. In spite of their different shapes, they are strictly equivalent from a mathematical standpoint. When studying the electronic structure of any complex, the choice of the proper z axis is crucial. It must take advantage of the geometrical structure as we shall see later.

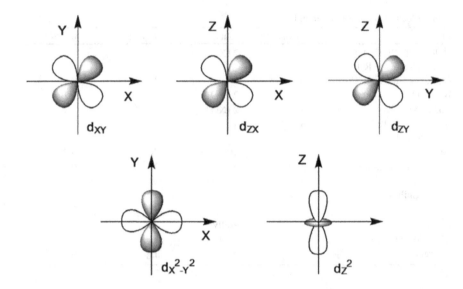

We shall see now how we build a M–H bond using these d orbitals. The choice of the z axis is obvious: it is the axis of the bond. The bond results from the overlap between the d_{z^2} orbital on the transition metal and the $1s$ orbital on hydrogen. Hydrogen is more electronegative (2.2) than any transition metal (between 1.22 for Zr and 1.75 for Ni), hence the $1s$ orbital is lower in energy (more stable) than the d_{z^2} orbital. The mixing of the two orbitals leads to a bonding orbital (positive overlap) which is mathematically represented by a linear combination of $1s$ with a small positive λ d_{z^2} component. This bonding orbital contains the two electrons of the bond and is mainly localized on hydrogen. Thus, this M–H species is a hydride (H δ^-). The picture is completed by an antibonding (destabilized) orbital (negative overlap) which is essentially the d_{z^2} orbital with a small negative λ $1s$ component, which is mainly localized on the metal and is empty. It is essential to remember that the strength of the bond (as measured by the stabilization δE of the bonding orbital with respect to the $1s$ orbital) is proportional to the square of the overlap S and inversely proportional to the energetic gap ΔE between d_{z^2} and $1s$.

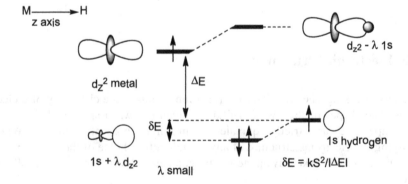

The picture is very similar for a M–L bond, except that the two electrons of the bond are provided by the lone pair.

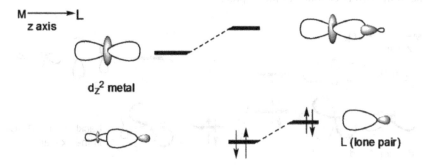

An additional complexity arises when the element carrying the lone pair also has empty orbitals of low energy and appropriate symmetry that can overlap with the d_{zx} (or d_{zy}) orbital of the metal and act as an acceptor of electronic density. This is the so-called backbonding that gives some double bond character to the M–L interaction. This happens with phosphorus where the acceptor orbitals are antisymmetric combinations of the P–R bond σ^* empty orbitals. This does not happen with nitrogen where the σ^* empty orbitals are too high in energy. This is the reason why phosphines can coordinate with both electron-rich (for example with OS = 0) and electron-poor metals (high-valent), whereas amines preferentially coordinate with high-valent metals. In the case of phosphines, the electron-rich metal can render some of its electronic density through its backbonding with phosphorus. It must be noticed that pyridines that display π^* empty orbitals are more similar to phosphines than to amines from this standpoint.

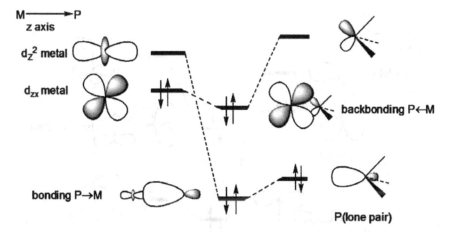

Carbon monoxide is a special case of interest. Besides its axial high-energy lone pair at carbon, it possesses two empty π_x^* and π_y^* orbitals that can overlap with the d_{zx} and d_{zy} orbitals at the metal. Thus, it can give two backbonds with the metal and is one of the most powerful acceptor ligand. The scheme only displays one of the two backbonds in the zx plane.

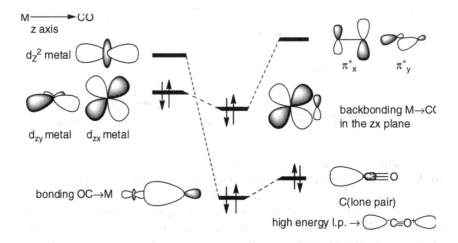

Bonding and backbonding also exist in the π complexes of alkenes with transition metals. The bond corresponds to a donation of electronic density from the π occupied orbital of the alkene to the empty d_{z^2} orbital of the metal. This d_{z^2} orbital has a cylindrical symmetry around the z axis. Thus, a rotation of the alkene around this axis does not modify the overlap between the π and d_{z^2} orbitals. This bond allows a free rotation of the alkene around the z axis.

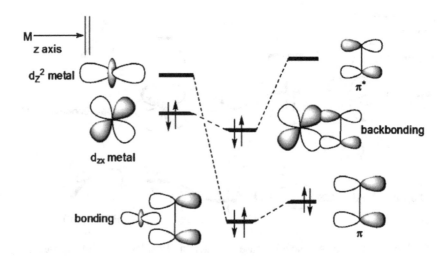

The situation is entirely different for the backbond. Both the d_{zx} and π^* orbitals are lying in the zx plane. A rotation of the alkene around the z axis breaks the backbond. Hence, measuring the rotation barrier of the alkene (for example, by NMR at variable temperature) gives the strength of the backbond. It can be noticed that the electronic density of the backbond lies outside of the C–M axes like the density of the C–C "banana" bonds of cyclopropanes. If the backbond is stronger than the bond (as this is the case with strong acceptors like C_2F_4), then the best representation of the complex is not a classical π bond but a metallacyclopropane.

1.4 Molecular Orbitals of Some Representative Complexes

A real complex includes a metal surrounded by several ligands. The electronic structure of this complex is dictated by its geometry as we shall see. For the purpose of simplicity, the ligands will be considered identical and will be represented by a lone pair (L). We shall start by the octahedral complex.

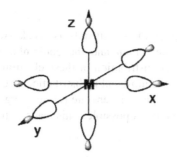

At low energy, we have the six lone pair orbitals of the ligands which, altogether, contain 12 electrons. At higher energy, we have the d orbitals of the metal. The three orbitals d_{xy}, d_{xz}, and d_{yz} do not interact with the lone pairs and are not destabilized. The d_{z^2} and $d_{x^2-y^2}$ are destabilized as a result of their negative overlaps with the lone pairs. Demonstrating that the destabilization of both orbitals is identical is not simple from a mathematical standpoint. But, from a physical standpoint, this is obvious. Indeed, our choice of the x, y, and z axis is arbitrary. If we exchange the z with the x axis, we must find an identical result since the complex has a physical reality independent from our choice of a virtual axis. Thus, we have a degenerate set of three orbitals, and at higher energy, a degenerate set of two orbitals. In order to reach the 18 electron configuration, we can put six electrons in d_{xy}, d_{xz}, and d_{yz}. We, thus, get a stable diamagnetic complex. It is labeled strong ligand field, low spin and is characterized by a large HOMO–LUMO gap.

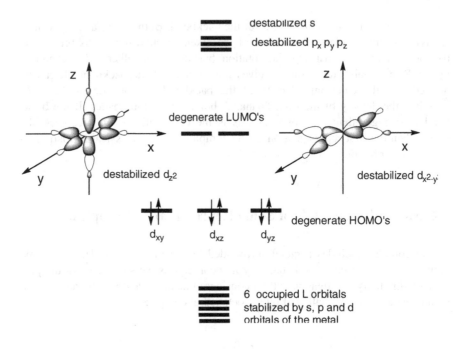

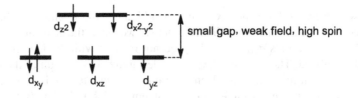

If the destabilizing effect of the ligands is weak (weak ligand field), the HOMO–LUMO gap becomes small and the coulombic repulsion between the electrons tends to change the distribution of these electrons among the d orbitals. The Hund's rule states that the most stable distribution of electrons among a set of degenerate (same energy) or quasi-degenerate (close energies) orbitals spreads the electrons onto as many orbitals as possible. This leads to the so-called weak field, high spin configuration.

The ligands can be ranked according to their ability to promote the high field, low spin configuration or the reverse. Since the HOMO–LUMO gap (destabilization of d_{z^2} and $d_{x^2-y^2}$) is inversely proportional to the energy gap between the metal d orbitals and the L orbitals, the higher the electronegativity of the metal-bonded atom in L, the lower the energy of the L orbitals, the higher the d/L gap, the lower the destabilization. Thus, a high electronegativity favors the high spin configuration. But other factors play a role. In fact, the following so-called spectrochemical series is observed:

High spin $I^- < Br^- < Cl^- < F^- < OH^- < OH_2 < NH_3 << CN^- < PR_3 < CO$ low spin

In the case of PR_3, the low-lying acceptor orbitals at P stabilize the d_{xy}, d_{xz} ,and d_{yz} orbitals, increase the HOMO–LUMO gap and favor the low-spin configuration. As an example, we can compare $[Fe(OH_2)_6]^{2+}$ (high spin) and $[Fe(CN)_6]^{2-}$ (low spin).

We shall now investigate the square planar ML_4 complexes. This type of complexes is quite important because many of the precatalysts have this structure (e.g., the Wilkinson catalyst $[RhCl(PPh_3)_3]$). ML_4 is obtained from ML_6 by removing the two ligands on the z axis. This removal sharply stabilizes d_{z^2} which is still, nevertheless, slightly destabilized by the interaction of its central torus with the ligands in the xy plane. The complex is stable with a 16-electron configuration : $8(L) + 6(d_{xy}, d_{xz}$,and $d_{yz}) + 2(d_{z^2})$. It has both a vacancy to accept

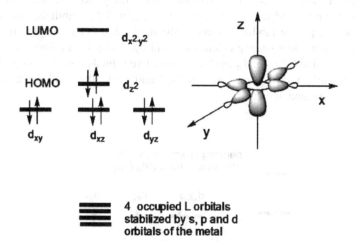

$\underline{\quad\quad\quad}$ 4 occupied L orbitals
stabilized by s, p and d
orbitals of the metal

an incoming ligand and a high-lying HOMO to fill an antibonding orbital and induce the breaking of the corresponding bond. This is a perfect configuration for a catalyst. Nevertheless, there is a problem. If we consider the initial interaction of such a complex with a dihydrogen molecule approaching along the z axis in the zx plane, we can see that the overlap of the empty σ^* of H_2 with d_{z^2} will be zero. Indeed, d_{z^2} is symmetrical whereas σ^* is dissymmetrical with respect to the zy plane. But the incoming H_2 repels the L_x and L_{-x} ligands and this distortion of the geometrical structure induces a change of the electronic structure of the complex. The effect of the bending of the complex around the y axis is shown below. The destabilization of the d_{z^2} orbital diminishes and its energy goes down whereas $\underline{d}xz$ is sharply destabilized and its energy goes up. At some point, d_{xz} becomes the HOMO and, since it has the good symmetry, it interacts with σ^*. The donation of electronic density from d_{xz} into the antibonding σ^* induces the breaking of the H–H bond.

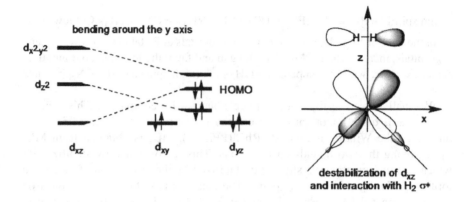

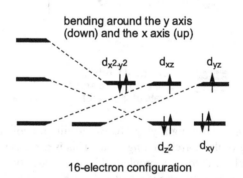

We have bent down the ML_4 complex around the y axis. If, in addition, we now bend it up around the x axis, d_{z^2} and $d_{x^2-y^2}$ will be stabilized further and d_{yz} will be sharply destabilized. Finally, the tetrahedral ML_4 complex has a set of two degenerate low energy orbitals d_{z^2} and d_{xy} and a set of three degenerate high energy orbitals $d_{x^2-y^2}$, d_{xz}, and d_{yz}. These five orbitals will be occupied for a 18-electron configuration. As a result of the Hund's rule, the 16-electron configuration will be paramagnetic.

Square planar (diamagnetic) and tetrahedral (paramagnetic) 16-electron ML_4 complexes are very close in energy. The complex $[Ni(PPh_2Et)_2Br_2]$ has been isolated in both forms, square-planar and tetrahedral; they are found to be in equilibrium in solution and are thus very close in energy.

We shall now briefly present the electronic configurations of some other types of complexes. The ML_5 complexes mainly exist in two varieties, the square pyramidal and the trigonal bipyramidal (TBP). The d blocks look as shown.

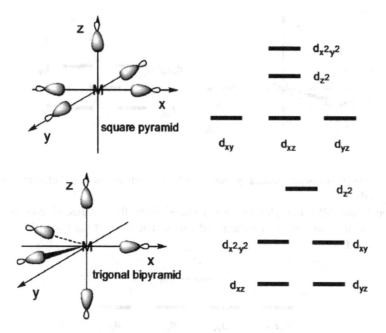

The TBP structure is highly fluxional (flexible), that is the axial ligands on the z axis can become equatorial (in the xy plane) through a geometrical distorsion known as the Berry pseudorotation.

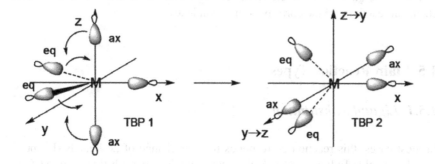

The y and z axis exchange their roles in this pseudorotation transforming TBP 1 into TBP 2. The L_{eq}–M–L_{eq} angle increases from 120 to 180°, while the L_{ax}–M–L_{ax} angle simultaneously decreases from 180 to 120°. This transformation needs a very small amount of energy. In the case of iron pentacarbonyl [$Fe(CO)_5$], the barrier is as low as 4 kJ mol^{-1} and the pseudorotation proceeds even at $-40\,°C$.

The trigonal planar ML_3 complex can be deduced from the TBP complex by simple removal of the two ligands on the z axis. This leads to a strong stabilization of d_{z^2} as shown.

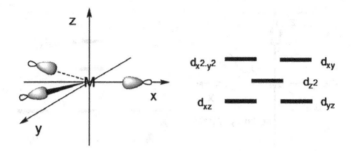

These complexes are generally stable with 16 electrons (d^{10} + 6 electrons from the ligands).

The linear ML_2 complex can be deduced from the octahedral complex by removal of the four ligands on the x and y axis. The sole orbital which is destabilized is the d_{z^2}.

These complexes generally have a 14 electron configuration and are highly reactive. They are common with copper, silver, and gold, e.g., $[CuPh_2]^-$, $[Ag(CO)_2]^+$, $[AuCl_2]^-$.

As can be guessed, these electronic configurations have a major impact on the chemical reactivity of the corresponding complexes.

1.5 Main Reaction Types

1.5.1 Ligand Substitution

In most cases, this reaction corresponds to an exchange of 2e ligands (L), but it can also involve 1e ligands (X). A classical example is the substitution of a CO by a phosphine in a metal carbonyl:

$$[Mo\,(CO)_6] + PR_3 \rightarrow [Mo\,(CO)_5\,(PR_3)] + CO$$

As a general rule, the coordination number, the number of electrons and the oxidation state of the central metal are preserved in this process. There are two basic mechanisms for this reaction, dissociative and associative. The dissociative mechanism is generally followed by the 18e complexes such as $[Mo(CO)_6]$ in the example before. The first step is the dissociation of one L ligand, giving a 16e complex. This is the rate-determining step. It is followed by a second, faster step in which the external ligand L' occupies the vacancy left by L in the first step. If this second step is sufficiently fast, no reorganization of the 16e complex can take place and the replacement of L by L' occurs with full retention of the metal stereochemistry. Otherwise, racemization can occur, especially if the 16e intermediate is pentacoordinate.

$$[L_nM–L] \xrightarrow{k} [L_nM] + L \quad \text{slow step}$$
$$[L_nM] + L' \longrightarrow [L_nM–L'] \quad \text{fast step}$$

The rate of the reaction obeys the following equation: $v = k\,[L_nM–L]$. An excess of the incoming ligand L' is useless. This mechanism resembles the S_N1 substitution in organic chemistry. The electrochemical reduction of the starting complex favors this mechanism: the resulting 19e complex dissociates more readily than the starting 18e complex. This process can be catalytic as shown below.

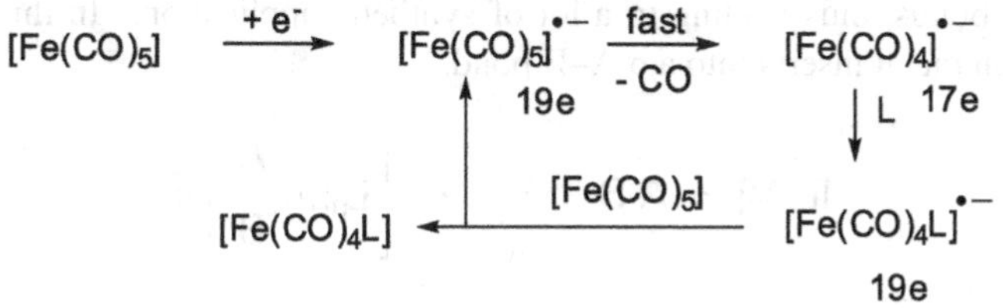

$$[Fe(CO)_5] \xrightarrow{+\,e^-} [Fe(CO)_5]^{\bullet-} \xrightarrow[-\,CO]{\text{fast}} [Fe(CO)_4]^{\bullet-}$$

$$19e \qquad\qquad 17e$$

$$[Fe(CO)_4L] \xleftarrow{\quad [Fe(CO)_5] \quad} [Fe(CO)_4L]^{\bullet-}$$

$$19e$$

When the replacement of a X ligand (X = Cl, Br, I) is required, the use of soluble Ag^+ or Tl^+ salts is helpful: it leads to the formation of insoluble AgX (or TlX) salts and the transformation of $[L_nM–X]$ into the 16e $[L_nM]^+$. An example is shown below. Contrary to the M–L bond which is purely covalent, the M–X bond has both covalent and ionic components and is generally stronger than the M–L bond. As a consequence, without silver or thallium salts, the substitution tends to involve the dissociation of the M–L bonds.

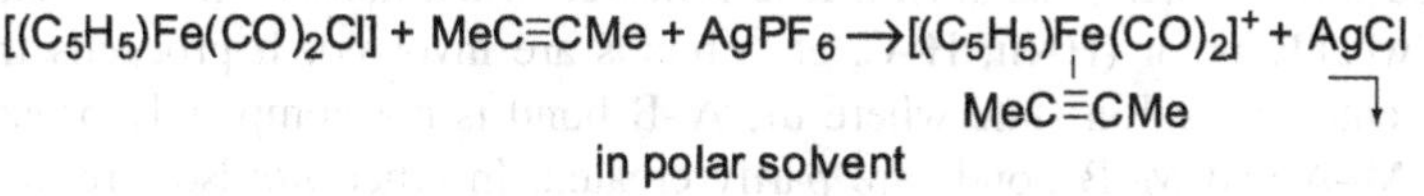

$$[(C_5H_5)Fe(CO)_2Cl] + MeC\equiv CMe + AgPF_6 \longrightarrow [(C_5H_5)Fe(CO)_2]^+ + AgCl$$

$$MeC\equiv CMe \quad\downarrow$$

in polar solvent

The associative mechanism is generally followed by electron deficient $(n < 18)$ complexes. The first step is the association of one L' ligand, giving a $(n + 2)e$ complex. This is the rate-determining step. It is followed by a second, faster step in which one L ligand is lost.

$$[L_nM] + L' \xrightarrow{k} [L_nM–L'] \quad \text{slow step}$$
$$[L_nM–L'] \longrightarrow [L_{n-1}ML'] + L \quad \text{fast step}$$

The rate of the reaction obeys the following equation: $v = k\,[L_nM][L']$. An excess of the incoming ligand L' accelerates the reaction. This mechanism resembles the S_N2 substitution in organic chemistry. The electrochemical oxidation (or even the oxidation by air) of the starting complex favors this mechanism: the resulting $(n-1)e$ complex is more reactive towards L' than the starting complex. This process can be catalytic as shown below.

$$[(C_5H_5)Mn(CO)_2(MeCN)] \xrightarrow{-e^-} [(C_5H_5)Mn(CO)_2(MeCN)]^{\cdot+} \xrightarrow{L} [(C_5H_5)Mn(CO)_2L]^{\cdot+}$$
$$\text{17e} \qquad\qquad\qquad \text{17e}$$

$$[(C_5H_5)Mn(CO)_2L] \longleftarrow$$
$$[(C_5H_5)Mn(CO)_2(MeCN)]$$

1.5.2 Oxidative Addition

One of the most useful characteristics of transition metals is their ability to activate strong bonds, thus leading to a lot of synthetic applications. In this process, the transition metal inserts into a σ A–B bond.

$$[L_nM] + \text{A-B} \longrightarrow \left[L_nM\diagup^{A}_{\diagdown B} \right]$$

In this reaction, the number of electrons, the oxidation state (hence the name) and the coordination number of the metal increase by two units. In most of the cases, the A–B bonds are H–H, H–X, H–Si, H–C, C–X... The metal is initially in a low oxidation state (mainly 0, 1, 2) and has a maximum of 16 electrons. It must be stressed that this process has only a limited counterpart in organic chemistry since carbon is essentially tetravalent. The insertion of carbene $[CH_2]$ (divalent carbon) into the O–H bond of carboxylic acids to give methyl esters is an example.

There are mainly four mechanisms for this type of reaction. The first one is the concerted three-center mechanism. It is followed when homoatomic (H–H, O–O, etc.) or weakly polar (H–Si, H–C, etc.) bonds are involved. It proceeds through a three-center transition state where the A–B bond is not completely broken and the two M–A and M–B bonds are partly created. In other words, it resembles a η^2 complex of a σ bond. Such complexes have been fully characterized with dihydrogen.

$$[L_nM] + \text{A-B} \xrightarrow{k} \left[L_nM\diagup^{A}_{\vdots \diagdown B} \right] \longrightarrow \left[L_nM\diagup^{A}_{\diagdown B} \right]$$

The initial product has the *cis* geometry. The reaction is not sensitive to the polarity of the solvent as expected for a concerted process. The kinetics are second order: $v = k\,[L_nM][\text{A-B}]$. The addition of dihydrogen onto the Vaska complex is a classical example:

square planar 16e   octahedral 18e

The second mechanism is similar to that of the S_N2 substitution in organic chemistry. Whenever the A–B bond is highly polar (H–X, R–X, etc.) and the metal is nucleophilic (for example, a square planar metal with a d_{z^2} lone pair), the metal tends to react with this bond at the electron poor center (A). The process is favored when B is a good leaving group (OTs $> I > $ Br $>$ Cl).

$$[L_nM] + A\text{-}B \xrightarrow{k} \left[L_nM-A \right]^+ + B^- \longrightarrow \left[L_nM\overset{A}{\underset{B}{}} \right]$$

The kinetics are second order as it is the case in the concerted mechanism but the final product can have the *cis* or *trans* geometry. As it is observed in any nucleophilic substitution, polar solvents are highly favorable. The reaction occurs with inversion of the stereochemistry at A. The reaction of the Vaska complex with methyl chloride is a classical example.

trans addition

At first sight, the radical mechanism looks very similar to the S_N2 mechanism.

$$[L_nM] + A\text{-}B \xrightarrow[\text{slow step}]{k} \left[L_nM-A \right]^{\bullet} + B^{\bullet} \longrightarrow \left[L_nM\overset{A}{\underset{B}{}} \right]$$

It is favored by the stability of the $B^{\bullet}$ radical. If it is a radical-chain process, it can be catalyzed by radical initiators (O_2, peroxides, AIBN, etc.) and quenched by inhibitors (bulky phenols). The problem is to distinguish between the S_N2 and the radical mechanisms. In some cases, this is possible because the $B^{\bullet}$ radical undergoes a very fast intramolecular rearrangement:

The ionic mechanism occurs in dissociating solvents with strong acids (A–H) or nucleophilic anions (A⁻). In the first case, the initial protonation step determines the rate of the reaction: $v = k\,[ML_n][H^+]$ whereas, in the second case, the rate is controlled by the association of the anion: $v = k\,[ML_n][A^-]$. The protonation requires a basic complex with low oxidation state and strong donor ligands whereas the anionic mechanism generally requires high oxidation states and positive charge. Finally, the most common oxidative addition pathways are the concerted and the S_N2 processes.

1.5.3 Reductive Elimination

This process generally allows the recovery of the organic product at the end of a catalytic cycle. It is the opposite of oxidative addition.

$$\left[L_nM{\overset{A}{\underset{B}{\diagup}}} \right] \longrightarrow [L_nM] + A\text{-}B$$

In order to be able to create the A–B bond, A and B must be *cis* in the coordination sphere. The oxidation state of the metal is reduced by two in the process, thus this reaction is favored by high oxidation states. When the reductive elimination does not spontaneously proceeds, it is possible to promote it by electrochemical oxidation of the central metal. It is also promoted by steric crowding in the coordination sphere and strong A–B bonds. The mechanism is generally concerted and takes place with retention of stereochemistry at A and B. This property is of crucial importance for asymmetric catalysis.

When the oxidation state of the metal preferentially changes by one unit, then bimetallic reductive eliminations can be observed.

$$[L_nM\text{–}A] + [B\text{–}ML_n] \longrightarrow A\text{-}B + 2\,[ML_n]$$

A classical example is the decomposition of tetracarbonylcobalt hydride:

$$2\left[CoH\,(CO)_4 \right] \rightarrow H_2 + \left[Co_2\,(CO)_8 \right]$$

Here, the initial step is the homolytic cleavage of the Co–H bond giving two radicals.

1.5.4 Oxidative Coupling and Reductive Decoupling

In this process, two π (alkenes, alkynes) or multiply bonded (carbenes, carbynes) ligands combine in the coordination sphere of the metal:

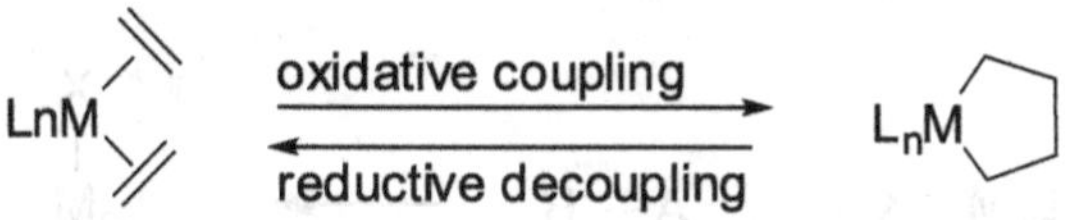

In the case of π ligands, the oxidation state of the metal increases by two units, whereas the electron count of the metal decreases by two units. The process is favored by electron-rich metals and by electron poor alkenes (low-lying LUMO). This situation favors the transfer of electrons from the metal into the π^* orbitals of the alkenes, leading to the breaking of the π bonds. One of the most classical and most useful examples of this reaction is the synthesis of zirconacyclopentadienes:

The coupling of one alkene with one carbene is the key step of the metathesis of alkenes that will be discussed later in this book.

1.5.5 Migratory Insertion, Elimination

In this process, an unsaturated 2-electron ligand A=B inserts into a M–X bond. Two types of insertion are observed, 1,1 and 1,2:

$$L_nM\text{—}X \ + \ A\text{=}B \ \longrightarrow \ L_nM\text{—}A\text{—}X \quad (1,1)$$
$$\underset{\displaystyle \| }{\phantom{L_nM\text{—}A}}$$
$$B$$

$$L_nM\text{—}X \ + \ A\text{=}B \ \longrightarrow \ L_nM\text{—}A\text{—}B\text{—}X \quad (1,2)$$

This is a two-step process and the type of insertion depends on the first step which involves the coordination of A=B to the metal:

$$
(1,1) \quad L_nM\!-\!X \;+\; A\!=\!B \quad\longrightarrow\quad L_nM\!-\!\underset{}{\overset{\displaystyle X}{|}}\!A\!=\!B
$$

$$
(1,2) \quad L_nM\!-\!X \;+\; A\!=\!B \quad\longrightarrow\quad L_nM\!-\!\underset{B}{\overset{\displaystyle X}{\underset{\|}{|}}}\!\!A
$$

Since X bears a negative charge, it can perform an intramolecular nucleophilic attack onto the A=B ligand; this second step is the migratory insertion, *stricto sensu*:

$$
(1,1) \text{ X attacks A:} \quad L_nM\!-\!\overset{X}{|}A\!=\!B \quad\longrightarrow\quad L_nM\!-\!\underset{B}{\overset{}{\underset{\|}{A}}}\!-\!X
$$

$$
(1,2) \text{ X attacks B:} \quad L_nM\!-\!\underset{A}{\overset{X\;\;\;B}{\underset{\|}{|}}} \quad\longrightarrow\quad L_nM\!-\!B\!-\!A\!-\!X
$$

In these second steps, the number of electrons on the metal decreases by two units, hence adding a 2e ligand will favor the process. In the (1,1) case, the oxidation state of A increases by two units, thus limiting the number of possibilities. In practice, only carbon monoxide, C=S, SO_2, carbenes, isonitriles R–N=C, NO can give such insertions.

$$
L_nM\!-\!X \;+\; CO \quad\longrightarrow\quad L_nM\!-\!\underset{O}{\overset{}{\underset{\|}{C}}}\!-\!X
$$
$$
C +2 \rightarrow +4
$$

$$
L_nM\!-\!X \;+\; SO_2 \quad\longrightarrow\quad L_nM\!-\!\underset{O}{\overset{O}{\underset{\|}{\overset{\|}{S}}}}\!-\!X
$$
$$
S +4 \rightarrow +6
$$

SO_2 can also give (1, 2) insertions. As expected, the case of NO is special:

$$
(C_5H_5)Co\!\!\underset{NO}{\overset{Me}{<}} \;\underset{(3e)}{\overset{(2e)}{\rightleftharpoons}}\; (C_5H_5)Co\!-\!\underset{O}{\overset{}{\underset{\|}{N}}}\!-\!Me \;\overset{L}{\longrightarrow}\; (C_5H_5)Co\!\!\underset{\underset{O}{\overset{\|}{N}}\!-\!Me}{\overset{L}{<}}
$$
$$
18\text{ e} \qquad\qquad\qquad 16\text{ e}
$$

Alkenes and alkynes give (1, 2) insertions:

$$\begin{array}{ccccc} \diagdown\!\!\!\diagup & & & \diagdown\!\!\!\diagup & \\ C{=}C & + & \left[L_nM{-}X\right] & + & C{=}C \\ \diagup\!\!\!\diagdown & & & \diagup\!\!\!\diagdown & \end{array} \longrightarrow L_nM{-}\overset{|}{\underset{|}{C}}{-}\overset{|}{\underset{|}{C}}{-}X$$

$$\left[L_nM{-}X\right] + {-}C{\equiv}C{-} \quad L_nM{-}X \longrightarrow \begin{array}{c}\diagdown\!\!\!\diagup\\C{=}C\\ \diagup\quad\diagdown\\ L_nM \quad X\end{array}$$

In the case of alkynes, the *cis* stereochemistry is characteristic of a process taking place inside the metal coordination sphere.

1.5.6 Nucleophilic Attack on Coordinated Ligand

If the metal is electron-poor (positive charge, high oxidation state, π-accepting coligands like CO), then a ligand bonded to the metal becomes more electrophilic. Nucleophilic attack onto this ligand becomes possible or becomes easier. Note that a metal complex can be electron-poor with a 18e configuration. It just means that the frontier orbitals of the complex are low in energy. If the product resulting from the nucleophilic attack stays in the coordination sphere of the metal, the reaction is called a nucleophilic addition. A typical example is given below.

Ethylene is not attacked by methylate anion when free.

If the product resulting from the nucleophilic attack leaves the coordination sphere of the metal, the reaction is called a nucleophilic abstraction. A typical example is given below.

Some special cases are particularly important. A first case is the attack of hydroxide ion onto a metal carbonyl, leading to a hydride with loss of CO_2.

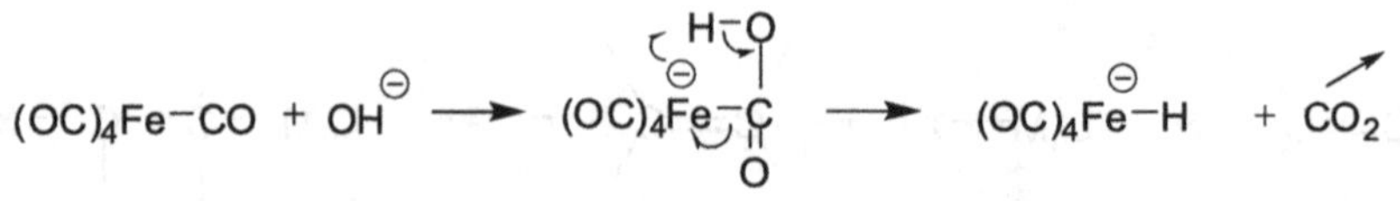

$$(OC)_4Fe-CO + OH^{\ominus} \longrightarrow (OC)_4Fe^{\ominus}-\overset{H-O}{\underset{O}{\overset{|}{C}}} \longrightarrow (OC)_4Fe-H + CO_2$$

The attack of a metal carbonyl by an amine oxide creates a vacancy at the metal, again with loss of CO_2.

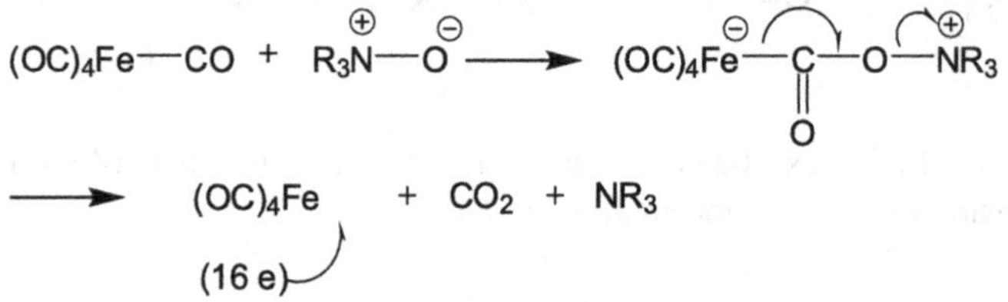

$$(OC)_4Fe-CO + R_3\overset{\oplus}{N}-\overset{\ominus}{O} \longrightarrow (OC)_4Fe^{\ominus}-\overset{|}{\underset{O}{C}}-O-\overset{\oplus}{N}R_3$$

$$\longrightarrow (OC)_4Fe + CO_2 + NR_3$$
$$(16\ e)$$

Of more general interest is the case of alkynes, alkenes and polyenes. In the case of alkynes, since the attack takes place at the opposite of the metal, the *trans* stereometry of the resulting alkenylmetal is opposite to that observed in the migratory insertion process. For the same reason, the stereochemistry of the products is *exo* with coordinated arenes.

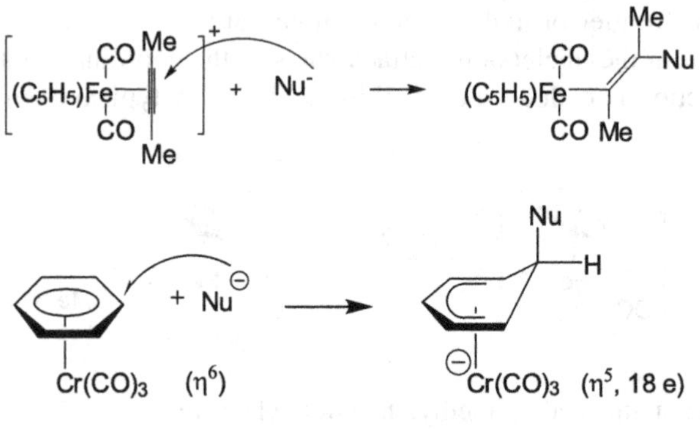

As a general rule, polyenes (η^{2n}) are more reactive than polyenyls (η^{2n+1}) due to the formal negative charge of the latter. The open species are more reactive than the cyclic species and, whenever possible, the *exo* attack takes place at the termini (steric reasons and higher localization of the LUMO).

1.5.7 Electrophilic Attack on Coordinated Ligand

Since the HOMO is normally highly localized at the metal, any electrophile tends to react at the metallic center. In order to promote the reaction at the ligand, it is necessary to block the attack at the metal by steric hindrance or to use a d^0 centre. As expected, this type of attack is favoured by electron-rich metals (negative charge, low oxidation state, σ-donating ligands). The following example with a d^0 centre is of synthetic interest.

retention at C

The vicinity of the highly polar Zr–C bond induces a polarization of the Br–Br bond. The retention of configuration at carbon has interesting uses in asymmetric synthesis.

1.6 Problems

I.1

Electron counts, oxidation states, and d^n configuration of:
$[ReH_9]^{2-}$, $TaMe_5$, $[(Ph_3P)_3Ru(\mu\text{-}Cl)_3Ru(PPh_3)_3]^+$
In this last case, do you think there is a metal–metal bond?

I.2

Electron counts, oxidation states, and d^n configuration of:
$MeReO_3$, $CpMn(CO)_3$, $[Re_2Cl_8]^{2-}$
In this last case, only terminal chlorines are present. Do you think there is a metal–metal bond? What multiplicity?

I.3

How many lone pairs are still available on chlorine in a covalent metal chloride Cl-M?

A complex is an oligomer of $Re(CO)_3Cl$. How can you write it with a $18e$ configuration at rhenium and no multiple bonds?

I.4

How many electrons are available for complexation on phosphorus in Ph_2P? What types of complexes can it give with a transition metal (call the metal M)?
Same questions for PhP.

I.5

What is the electronic configuration of $(OC)_3Co(NO)$? Is the Co-NO unit bent or linear? What is the oxidation state of cobalt?

I.6

What is the electronic configuration of nickelocene Cp_2Ni? Explain how it can react with a 2e ligand L and IMe to give the diamagnetic complex CpNi(L) I. What is the cyclopentadienyl (Cp) by-product? What is the product if only L is added?

I.7

Ph_2P-H reacts with $W(CO)_6$ to give ultimately the binuclear complex $(OC)_4W(\mu H)(\mu Ph_2P)W(CO)_4$. Give the developed formula of the complex. What is the electronic configuration and the oxidation state of tungsten? Is there a metal–metal bond? (μ means bridging). Propose a logical mechanism for the formation of the binuclear product. Each elementary step must be detailed.

I.8

What are the two possible choices for the oxidation state and number of electrons of iron in the nitroprusside ion $[Fe(CN)_5NO]^{2-}$? What is the most likely? Is Fe–N–O linear or bent? Is the salt dia—or paramagnetic?

I.9

The following reaction proceeds in two steps:

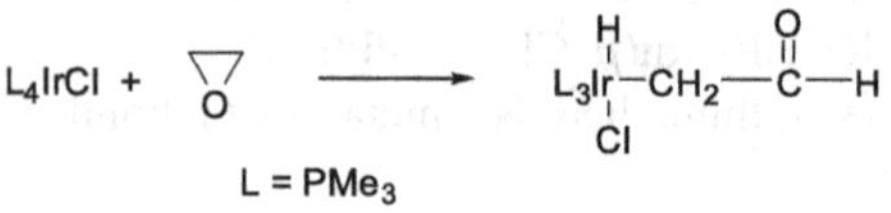

$$L_4IrCl + \triangledown\!\!-O \longrightarrow L_3\overset{H}{\underset{Cl}{Ir}}-CH_2-\overset{O}{\overset{\|}{C}}-H$$

$$L = PMe_3$$

What is the oxidation state of iridium before and after the reaction? What are the two elementary reactions?

I.10

$Ni(CO)_4$ and $Co(linear-NO)(CO)_3$ are isoelectronic, Why? Both have the same tetrahedral structure. The reaction of PPh_3 gives a monosubstituted product in both cases. What is the formula of this product in the cobalt case? For nickel, the mechanism of the substitution is dissociative, for cobalt it is associative. Explain this difference.

I.11

The reaction of $PhC \equiv CPh$ with $Fe_2(CO)_9$ gives, among other products, 2,3,4,5-tetraphenylcyclopentadienone. Propose a mechanism for the formation of this product.

What kind of complexes with transition metals can be formed with this cyclopentadienone?

I.12

A terminal alkyne can be coupled with an alkene in the presence of a ruthenium catalyst:

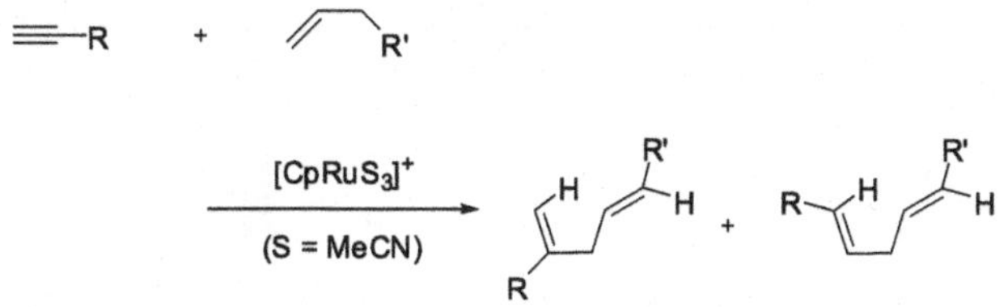

Propose a mechanism explaining the two regioisomers and the stereochemistries.

References

1. Jean Y (2005) Molecular orbitals of transition metal complexes, Oxford University Press, Oxford
2. Elschenbroich C (2006) Organometallics, 3rd edn. Wiley-VCH, Weinheim
3. Astruc D (2007) Organometallic chemistry and catalysis. Springer, Berlin
4. Robert H (2009) Crabtree, the organometallic chemistry of the transition metals, 5th edn. Wiley, Hoboken

Chapter 2
Main Types of Organometallic Derivatives

Abstract This chapter describes the various functional derivatives that form the backbone of transition metal chemistry. In each case, the coordination modes of the involved ligand are presented, then the main synthetic routes, the reactivity, and the most useful analytical techniques are described. For metal hydrides, the more specific points concern the η^2-H_2 complexes and the influence of spectator ligands on the acidity–basicity of hydrides in solution. For metal carbonyls, the high lability of the structures is stressed with its consequences for their analysis by IR or ^{13}C NMR. For metal alkyls or aryls, the various decomposition pathways are discussed with a special emphasis on the β-H elimination of importance for polymerization catalysis. The section is completed by a presentation of the uses of the zirconium–carbon bond in organic synthesis. The section on metal carbenes starts by a thorough discussion of the factors favoring the singlet or triplet ground states in free carbenes. Their complexation by transition metals leads to electrophilic (Fischer) or nucleophilic (Schrock) complexes with very different reactivities. Their role in the metathesis of alkenes is highlighted. A similar presentation of metal carbynes and their role in the metathesis of alkynes completes this section. The final section describes some specific π-complexes, η^4-diene-iron-tricarbonyls, ferrocene and η^6-arene-chromium-tricarbonyls which are widely used in organic synthesis.

Keywords Metal hydrides • Metal carbonyls • Metal alkyls • Metal carbenes • π-complexes

2.1 Metal Hydrides

For a long time, the existence of stable metal hydrides was controversial because the M–H bond is highly reactive and was difficult to detect before the emergence of proton NMR spectroscopy. The main coordination modes of hydrogen are indicated below.

F. Mathey, *Transition Metal Organometallic Chemistry*,
SpringerBriefs in Molecular Science, DOI: 10.1007/978-981-4451-09-3_2,
© The Author(s) 2013

Terminal H   (η^1)   H−MLn

Bridging H   (μ_2)

$$L_nM \xrightarrow{\quad H \quad} MLn$$

Bridging H   (μ_3)

$$LnM \underset{\underset{Ln}{M}}{\overset{H}{-\!\!\!-}} MLn$$

The bridging hydrides are considered as protonated M–M bonds (μ_2) or protonated clusters (μ_3). The M–H–M bond is always bent: M–H–M angle between ca. 80 and 120 °. These bridging species are a little bit delicate to handle for the electron counts. The best way is to count the number of electrons of the ML_n units, count the M–M bond if existing, then add the electrons of H and other bridging ligands.

$$Cp(OC)_2Mo \underset{H}{\overset{Me\ \ Me}{\underset{\displaystyle P}{\diamond}}} Mo(CO)_2Cp$$

$$\left\{ \begin{array}{ll} Mo & 6\ e \\ Me_2P & 3\ e \\ CO & 2\ e \\ Cp & 5\ e \\ H & 1\ e \end{array} \right.$$

Here, each metal unit has 15e, the Mo–Mo bond counts for 1, the two bridging ligands share their 4e between the two molybdenum atoms. The other important case concerns the η^2-H_2 complexes discovered by Kubas in 1984 [1]. The bonding can be decomposed into a donation from the H–H σ bond to the d_{z^2} empty orbital of the metal and a back donation from the d_{xz} orbital of the metal to the σ* antibonding orbital of H–H. In order to avoid the cleavage of H–H, this backbonding must be suppressed or minimized. This can be achieved when using metal centers with low d electron counts and electron-accepting ancillary ligands like CO, etc. The existence of the H–H bond in these complexes is established by IR spectroscopy (the stretching frequency of η^2-H–H corresponds to a band around 2700 cm^{-1}), by proton NMR (replacing H–H by H–D and measuring the H–D coupling), and by neutron diffraction.

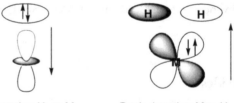

Donation H_2 to M   Back-donation M to H_2

The synthesis of hydrides is classical in most cases: oxidative addition of H_2, reduction of M–Cl by LiAlH$_4$ or other source of H$^-$, protonation, hydrogenolysis of M–M bonds, etc. A good example is the synthesis of the Schwartz reagent which is useful for the hydrozirconation of alkenes:

$$Cp_2ZrCl_2 \xrightarrow{LiAlH_4} Cp_2Zr(H)Cl + Cp_2ZrH_2 \xrightarrow{CH_2Cl_2} Cp_2Zr(H)Cl$$

Apart from these classical routes, it is also possible to get hydrides using the so-called β-H elimination. It operates with metal alkyls and metal alkoxides when β-H is available, and also with metal formates and metal hydroxycarbonyls.

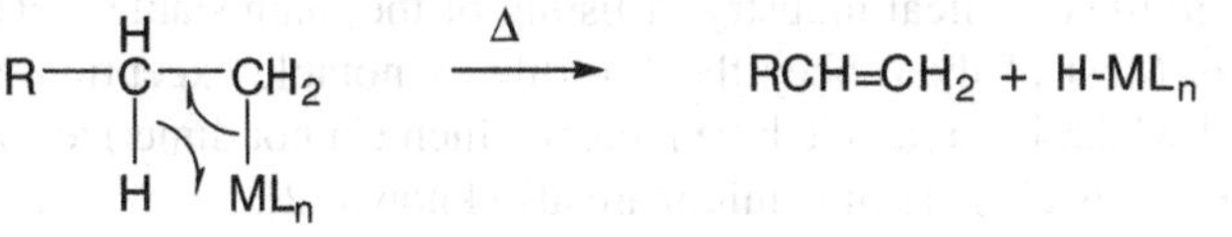

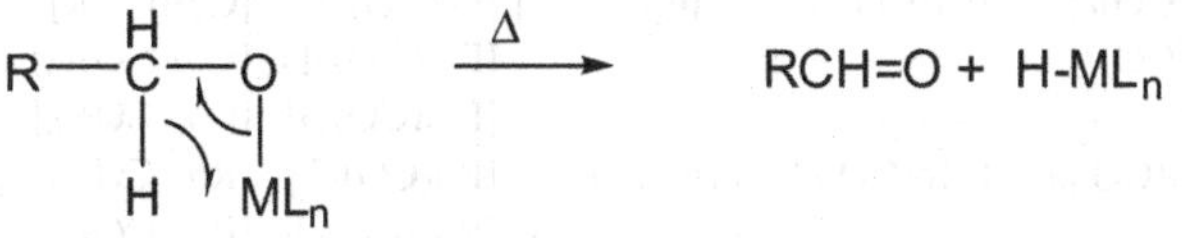

$$R-\overset{\overset{\displaystyle H}{|}}{\underset{\underset{\displaystyle ML_n}{|}}{C}}-CH_2 \quad \xrightarrow{\Delta} \quad RCH=CH_2 + H\text{-}ML_n$$

$$R-\overset{\overset{\displaystyle H}{|}}{\underset{\underset{\displaystyle ML_n}{|}}{C}}-O \quad \xrightarrow{\Delta} \quad RCH=O + H\text{-}ML_n$$

This route from alkoxides works only with nonoxophilic metals, mainly from the platinum group (Ru, Rh, Pd, Ir, Pt), for example:

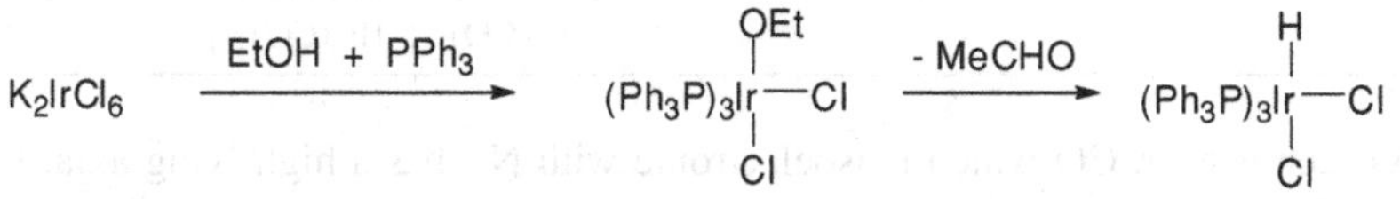

$$K_2IrCl_6 \xrightarrow{EtOH \ + \ PPh_3} (Ph_3P)_3\overset{\overset{\displaystyle OEt}{|}}{\underset{\underset{\displaystyle Cl}{|}}{Ir}}-Cl \xrightarrow{-\ MeCHO} (Ph_3P)_3\overset{\overset{\displaystyle H}{|}}{\underset{\underset{\displaystyle Cl}{|}}{Ir}}-Cl$$

As already mentioned, the best method for detecting the M–H bond is proton NMR spectroscopy. The hydride resonance appears between 0 and −50 ppm (Me_4Si), in a range which is empty for organic groups. The only exception is for d^0 and d^{10} metal hydrides which resonate at low fields (δ positive). The M–H bond is relatively strong; the bond dissociation energy (BDE) varies between 37 and 65 kcal mol^{-1}. Generally, these hydrides are poorly soluble in water and not compatible with this solvent. Nevertheless, [HCo(CO)$_4$] is an acid as strong as sulfuric acid whereas [HRe(C$_5$H$_5$)$_2$] is a base comparable to ammonia. It must be noted, however, that [HCo(CO)$_4$] is a genuine hydride in the gas phase (−0.75 e on H). More general and precise data can be measured in acetonitrile as the solvent:

Some pKa in CH$_3$CN:

[HCr(CO)$_3$Cp] 13.3 [HMo(CO)$_3$Cp] 13.9 [HW(CO)$_3$Cp] 16.1 As can be seen, the hydride character increases with the atomic weight of the metal.

[HCo(CO)$_4$] 8.4 [HCo(CO)$_3$(PPh$_3$)] 15.4

The influence of the other ligands on the metal is enormous. H is more basic when L is a better donor.

Some of the most useful reactions of metal hydrides are given hereafter:

$$M-H + CCl_4 \rightarrow M-Cl + CHCl_3$$
$$M-H + CO_2 \rightarrow M-C(O)OH$$
$$M-H + H_2C=CH_2 \rightarrow M-CH_2-CH_3$$
$$M-H + HC \equiv CH \rightarrow M-CH=CH_2$$

The third reaction is the reverse of the β-H elimination. Metal formyls [M–CHO], which are the primary products of the insertion of CO into the M–H bonds, are unstable in most cases but are important catalytic intermediates.

2.2 Metal Carbonyls

Metal carbonyls have been known since the end of the nineteenth century and are widely used in the chemical industry. A listing of the main stable metal carbonyls is given here. Most of them obey the 18e rule. A notable exception is vanadium hexacarbonyl which is a reactive blue radical which cannot dimerize for steric reasons. Several heavy clusters of osmium are also known [2].

$[Ti(CO)_6]^{2-}$	$[V(CO)_6]^-$ $[V(CO)_6]$	$[Cr(CO)_6]$	$[Mn_2(CO)_{10}]$	$[Fe(CO)_5]$ $[Fe_2(CO)_9]$ $[Fe_3(CO)_{12}]$	$[Co_2(CO)_8]$ $[Co_4(CO)_{12}]$ $[Co_6(CO)_{16}]$	$[Ni(CO)_4]$
$[Zr(CO)_6]^{2-}$	$[Nb(CO)_6]^-$	$[Mo(CO)_6]$	$[Tc_2(CO)_{10}]$ $[Tc_3(CO)_{12}]$	$[Ru(CO)_5]$ $[Ru_2(CO)_9]$ $[Ru_3(CO)_{12}]$	$[Rh_2(CO)_8]$ $[Rh_4(CO)_{12}]$ $[Rh_6(CO)_{16}]$	$[Pd(CO)_4]^{2+}$
$[Hf(CO)_6]^{2-}$	$[Ta(CO)_6]^-$	$[W(CO)_6]$	$[Re_2(CO)_{10}]$	$[Os(CO)_5]$ $[Os_2(CO)_9]$ $[Os_3(CO)_{12}]$	$[Ir_2(CO)_8]$ $[Ir_4(CO)_{12}]$ $[Ir_6(CO)_{16}]$	$[Pt(CO)_4]^{2+}$

As seen earlier, CO which is isoelectronic with N_2, has a high-lying axial lone pair at C which plays the major role in the coordination with a transition metal. The coordination mode can be η^1, μ_2, or μ_3

In some very rare instances, the coordination can also involve the two degenerate π bonds to give 4e and 6e complexes. Metal carbonyls are highly fluxional and can adopt several structures in the solid state and in solution. The classical example is $[Co_2(CO)_8]$ which displays bridging COs in the solid state but none in solution. This difference can be detected by IR spectroscopy.

solid state

solution

2069, 2055, 2032 cm^{-1}

solid state: ν(CO) 2071, 2044, 2042, 1866, 1857 cm^{-1}

bridging COs

In the same vein, the axial and equatorial COs of the trigonal bipyramidal $[Fe(CO)_5]$ cannot be distinguished by ^{13}C NMR in solution at room temperature due to their rapid interchange by Berry pseudorotation as discussed earlier. Their separation occurs at -38 °C in the solid state.

As seen previously, CO has two orthogonal π^* accepting orbitals, hence the existence of a strong backbonding in metal carbonyls. This backbonding has several consequences: (1) the C–O bond is lengthened: it increases from 1.128 Å in free CO to 1.13–1.18 Å in metal carbonyls; (2) the weakening of the CO bond can be monitored by IR spectroscopy since the CO stretching frequency is proportional to the square root of the force constant. In free CO, $\nu(CO)$ 2143 cm^{-1}, k = 19.8 mdyne/Å, in M–CO (terminal) $\nu(CO)$ 1900–2100 cm^{-1}, k = 17–18 mdyne/Å; (3) the M–C bond is strengthened and becomes shorter: in [Me–Mn(CO)$_5$], Me–Mn 2.185 Å, Mn–CO 1.80 Å. Overall, CO is one of the strongest π-acceptor ligand; the order is: NO > CO > RNC $\approx$ PF$_3$ > PCl$_3$ > P(OR)$_3$ > PR$_3$ $\approx$ SR$_2$ > RCN > RNH$_2$ $\approx$ OR$_2$.

The M–CO bond is relatively weak but its strength varies significantly according to the metal: in [Fe(CO)$_5$], the Fe–C bond strength is 27.7 kcal mol^{-1}, in [Cr(CO)$_6$], the Cr–C bond strength is 37 kcal mol^{-1}, in [Mo(CO)$_6$], the Mo–C bond strength is 40 kcal mol^{-1}, and in [W(CO)$_6$], the W–C bond strength is 46 kcal mol^{-1}. As a consequence, metal carbonyls are ideal substrates for substitution reactions. Also, in many cases, clusterization takes place easily by partial loss of CO under heating or irradiation by UV light:

$$[Co_2(CO)_8] \xrightarrow{70\,°C} [Co(CO)_3] \rightarrow [Co_4(CO)_{12}]$$

$$[Fe(CO)_5] \xrightarrow{h\nu} [Fe_2(CO)_9]$$

The best analytical tool to detect metal carbonyls is IR spectroscopy. The CO stretching frequencies are in the range 1900–2100 cm^{-1} for terminal COs and 1700–1850 cm^{-1} for μ_2CO's. The technique is both very sensitive and very fast (time constants in the range of 10^{-15} s.), so that it takes a snapshot of the molecule before it can fluctuate. In so doing, it gives information on the local symmetry of the complex. Here are some indications on the correlation between symmetry and CO stretching modes.

Complex	Symmetry	$\nu(CO)$ modes
[Ni(CO)$_2$L$_2$]	C_{2v}	2
Trans [RuX$_2$(CO)$_2$L$_2$]	D_{2h}	2
[CpMn(CO)$_3$]	C_{3v}	2
[Fe(CO)$_3$L$_2$] (axial L)	D_{3h}	1
[XMn(CO)$_5$]	C_{4v}	3
Trans [Mo(CO)$_4$L$_2$]	D_{4h}	1
[Ni(CO)$_4$]	T_d	1
[Cr(CO)$_6$]	O_h	1

It is possible to understand this correlation between the symmetry of a complex and the number of visible CO stretching modes on a simple example. Let us consider the *cis* and *trans* octahedral complexes [M(CO)$_2$L$_4$]. The two COs can vibrate in phase or out of phase. Thus, two bands would be expected in both cases. But in the *trans* case, the in phase vibration induces no change in the dipole moment of the molecule and cannot be detected by IR.

The other technique for the detection of metal carbonyls is ^{13}C NMR spectroscopy. The carbonyl resonances appear in the range 190–230 ppm. But this spectroscopy is slow (time constants in the range of 10^{-1} s.) and gives no reliable information on the local symmetry under standard conditions. For example, at room temperature, the trigonal bipyramidal [Fe(CO)$_5$] gives only one CO resonance at 210 ppm.

2.3 Metal Alkyls and Aryls

The metal–carbon σ bond is moderately strong (30–65 kcal/mol) but, in general, kinetically unstable because several pathways exist for its decomposition. As a general rule, the orders of thermodynamic stability are M–H > M–R, M–Ar > M–R, and M–Rf(perfluoroalkyl) > M–R. The most efficient decomposition pathway is the β-H elimination:

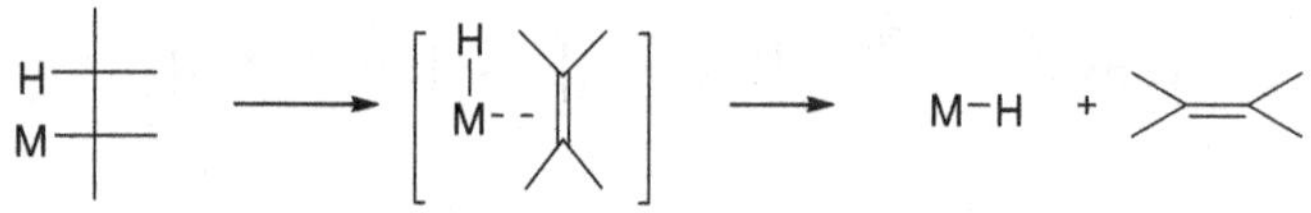

In order to confer some kinetic stability to the metal–alkyl bond, it is necessary to block this decomposition pathway. There are several possibilities:

(1) To use alkyl groups without β-H: CH$_3$, CH$_2$Ph, CH$_2$CMe$_3$, CH$_2$SiMe$_3$, CH$_2$CF$_3$, etc.
(2) To block the approach of β-H to the metal by steric hindrance or geometrical constraint (R = –C≡C–H linear)
(3) To block the formation of the alkene (when R is 1-norbornyl, the alkene cannot form at the bridgehead according to the Bredt rule)
(4) To block the formation of the alkene–hydride intermediate by using a stable 18e complex that cannot generate the needed vacancy by release of one of its L ligand. For example, in [CpFe(CO)$_2$Et], both Cp and CO are strongly bonded to iron due to a complementary push–pull electronic effect and the Fe–Et bond is stable.

In some cases, even when the β-H elimination appears possible (the M...H interaction can be detected either by spectroscopy or neutron diffraction), the decomposition does not proceed further and the metal alkyl becomes reasonably stable. This type of β-H interacting with the metal is called agostic. The species can be viewed as a loose η^2 complex of the C–H bond.

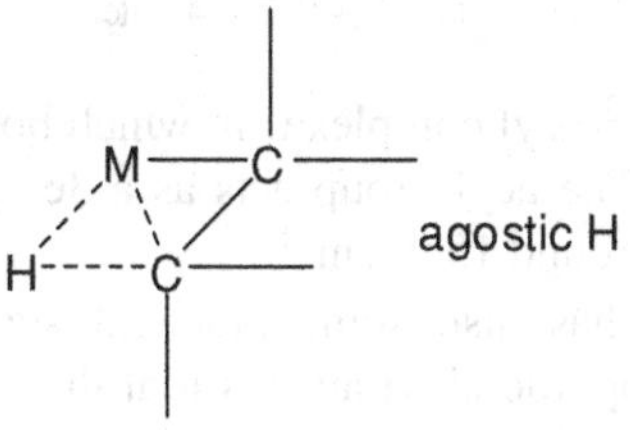

agostic H

The orbital scheme is similar to that of the η^2-H_2 complexes. In order to stabilize this kind of complexes, it is necessary to suppress the back donation of electrons from the metal to the σ* orbital of the C–H bond. This can be achieved by using d^0 metal centers. This is the reason why metals like Ti^{+4} and Zr^{+4} are used in olefin polymerization catalysis and in organic synthesis because both need reasonably stable alkyl complexes as we shall see later.

The other decomposition pathways of the M–C bond are the reductive elimination involving alkyl or aryl groups and the α-H elimination that leads to carbene complexes as discussed later.

The spectroscopic detection of the M–C bond is not as easy as it is for hydrides and carbonyls. With NMR-active nuclei (spin ½), the ^1J (M–C) coupling is useful. The list of convenient metals is given here.

Spin 1/2	Isotopic abundance
^{103}Rh	100
^{183}W	14
^{187}Os	1.6
^{195}Pt	34

The most useful chemistry of the M–C bonds deals with the insertion of small molecules. Among them, carbon monoxide deserves special attention. The normal scheme is depicted hereafter.

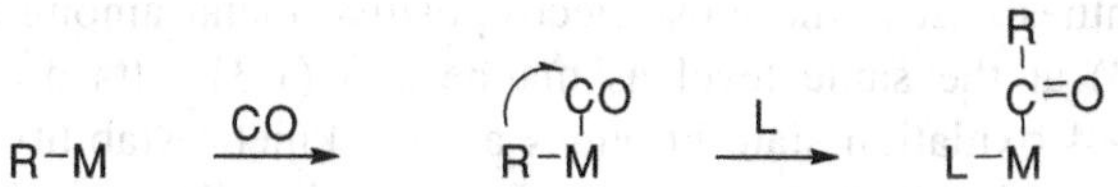

$$R-M \xrightarrow{\ CO\ } \overset{CO}{R-M} \xrightarrow{\ L\ } \overset{\overset{R}{|}}{\underset{L-M}{C=O}}$$

The CO stretching frequencies of the resulting acylmetal complexes appear around 1650 cm^{-1} like for an organic ketone. The situation is different with oxophilic metals from the left of the periodic table:

$$R-M \xrightarrow{\ CO\ } R-M\overset{\displaystyle \frown}{\underset{}{\text{CO}}} \longrightarrow M\overset{\displaystyle \overset{R}{\underset{\displaystyle C}{|}}}{\underset{O}{\triangle}}$$

M = Ti(+4), Zr(+4), etc.

Here, the products are η^2-acyl complexes in which both carbon and oxygen are coordinated to the metal. The acyl group acts as a 3e ligand. The CO stretching frequency is much lower, around 1550 cm^{-1}.

The insertion of CO_2 has also some practical significance. CO_2 gives an η^2-complex and the nucleophilic alkyl attacks it at the positive carbon in the vast majority of cases:

$$R-M \xrightarrow{\ CO_2\ } R-M\ \overset{\displaystyle O}{\underset{\displaystyle}{\text{C}-O}} \longrightarrow R-\overset{\displaystyle O}{\underset{\displaystyle}{C}}-O-M$$

The attack at oxygen giving M–C(O)OR is observed in some rare cases with nonoxophilic metals.

The case of SO_2 is more subtle because there is still a lone pair at sulfur that is liable to give $M \leftarrow SO_2$ complexes. The other coordination mode is η^2 and the final products are either the metallasulfones $M-S(O)_2R$ with thiophilic metals (nickel for example) or the sulfinates $M-OS(O)R$ with oxophilic metals (titanium for example). An equilibrium between both types is sometimes observed.

$$R-M \xrightarrow{\ SO_2\ } \begin{cases} R-M\ \overset{O\!\!\diagdown\,S\,\diagup\!\!O}{} \longrightarrow R-\overset{\displaystyle O}{\underset{\displaystyle O}{S}}-M \\[3em] R-M\ \overset{O\!\!\diagdown\,S-O}{} \longrightarrow R-\overset{\displaystyle O}{S}-O-M \end{cases}$$

2.4 The Zirconium–Carbon Bond in Organic Synthesis

Zirconium combines two features that explain its special role in stoichiometric organic synthesis. It is the most electropositive metal among the transition elements (1.33) at the same level as magnesium (1.31). Its d^0 configuration in its normal +4 oxidation state provides a good kinetic stability to the zirconium–carbon bond by disfavoring the β-H elimination. The Zr–C bond is classically obtained by hydrozirconation of alkenes. The zirconium hydride used is generally the easily accessible Schwartz reagent whose synthesis has been mentioned in the section on hydrides. When using a mixture of internal and terminal alkenes, only the linear alkylzirconium species are obtained, zirconium preferring the terminal position for steric and electronic reasons. With alkynes, the addition of Zr–H is *cis* and zirconium occupies the less hindered position. With conjugated dienes, only the 1,2 addition is observed. The zirconium–carbon bond easily inserts carbon monoxide and isonitriles but does not react with alkyl halides. The cleavage is easily obtained by reaction with acids (formation of RH), bromine, iodine, *N*-chlororosuccinimide (formation of RX) and hydrogen peroxide (formation of ROH). Zirconium can be replaced by another metal using the reaction with metal halides derived from less electropositive metals (Al, Cu, Ni).

Another reagent of choice for the synthesis of the zirconium–carbon bond is the transient zirconocene [ZrCp$_2$] obtained by reaction of butyllithium with the stable zirconocene dichloride Cp$_2$ZrCl$_2$. The intermediate Cp$_2$ZrBu$_2$ loses butane and yields the η^2 complex of zirconocene with butene. This zirconocene reacts with two molecules of alkynes to give a zirconacyclopentadiene. In turn, this zirconacyclopentadiene can serve to prepare boroles, stannoles, siloles, phospholes, arsoles, etc. by metathesis with the appropriate dihalides.

M = BR, SiR$_2$, SnR$_2$, PR, AsR

Another topic of interest concerns the stabilization of benzyne. This species has been transiently generated by various techniques in organic synthesis. It gives a stable η^2 complex with zirconocene which is obtained by thermolysis of Cp_2ZrPh_2 at 70 °C (loss of PhH) and further stabilized by complexation with Me_3P. This complex is, in fact, a zirconacyclopropene derived from $Zr(+4)$. The complexed CC bond has the typical length of a C=C double bond (1.36 Å). It displays a rich chemistry as shown below.

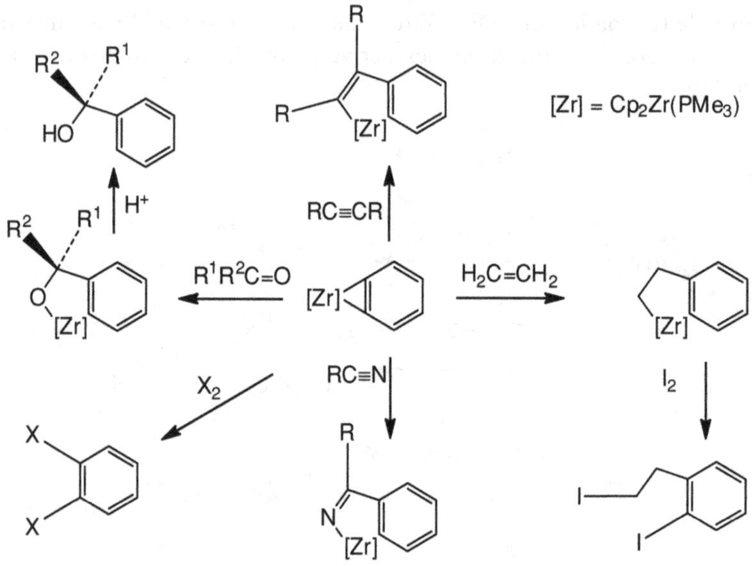

2.5 Metal Carbenes

Metal carbene complexes are schematically divided into two classes, the so-called Fischer carbenes, discovered by Fischer in 1964, and the Schrock carbenes discovered 10 years later by Schrock. The Fischer carbenes are characterized by an electrophilic carbon, whereas the Schrock carbenes are nucleophilic. Both types are conventionally represented as having an M=C double bond but this representation is misleading. In fact, the best representation of the Fischer carbenes would display a single dative bond $R_2C \rightarrow M$, the Schrock carbenes being the only ones with a genuine $R_2C=M$ double bond. In the first case, the carbene acts as an L ligand and the oxidation state of the metal is 0, whereas in the second case, the carbene is equivalent to X_2 and the oxidation state of the metal is +2.

Before discussing the electronic structures of Fischer and Schrock carbenes, it is necessary to discuss the electronic structure of free carbenes themselves. These 6e species display two frontier orbitals (σ and p) which are shown in the scheme and correspond to the σ and π bond of an alkene which results from the combination of two carbenes.

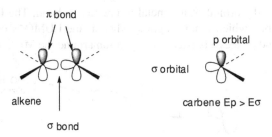

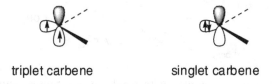

The two electrons of carbon that are not involved in its bonds can occupy the lower σ orbital to give a singlet. Alternatively, each electron can occupy a different orbital to give a triplet. The energies of these two orbitals are close, so the Hund's rule applies (see the section on ML_6 complexes) and states that the triplet is more stable than the singlet because it minimizes the interelectronic coulombic repulsion.

In fact, the triplet state of $[CH_2]$ is more stable than the singlet state by ca. 10 kcal mol^{-1}. The situation changes completely if one (or two) hydrogen is replaced by a lone pair substituent like oxygen or nitrogen. The singlet is strongly stabilized by interaction between the lone pair of the substituent and the vacant p orbital of the carbene whereas no sizeable stabilization occurs in the triplet (one electron is in a destabilized orbital). As a result, the singlet becomes the ground state.

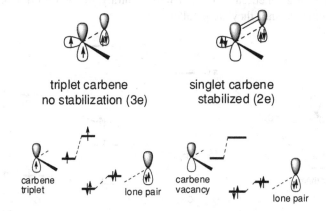

The Schrock carbenes correspond to the combination of a triplet carbene with a triplet metallic moiety as shown. The z axis is the axis of the double bond and the p orbital of the carbene is in the xz plane. The σ orbital of the carbene overlaps with the d_{z^2} of the metal to create a σ bond whereas the p orbital of the carbene overlaps with the d_{xz} orbital of the metal to create a π bond. The HOMO is centered on the carbon which is nucleophilic whereas the LUMO is centered on the metal. The π bond creates a barrier to the rotation around the M=C bond.

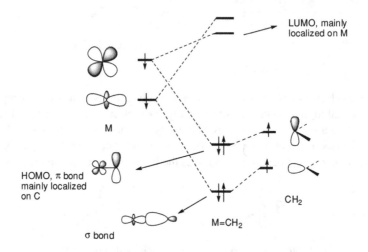

The situation is more complex with singlet heteroatom-substituted carbenes. Since the lone pair on the heteroatom is involved, four electrons must be taken into account. The case of [CHOH] is illustrated below. The singlet is more stable than the triplet by 17 kcal mol^{-1}. From an orbital standpoint, water and carbene are similar but the oxygen has two lone pairs, one σ and one π. Since C is less electronegative than O, its frontier orbitals are higher in energy e similar orbitals of O and, thus, are destabilized. The overlap between the two σ lone pairs at C and O is minimal for geometrical reasons. On the contrary, the two π orbitals at C and O strongly interact since they are parallel.

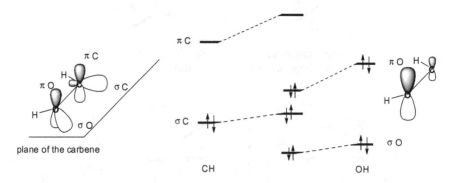

The Fischer carbenes typically involve the combination of a singlet carbene like [CHOH] with a singlet metal moiety. Since σ O is too low in energy to significantly interact with the metal orbitals, it suffices to consider σ C, π O, and π C when building M=CHOH.

σ C is stabilized by interaction with d_{z^2}. This orbital corresponds to the dative bond M $\leftarrow$ C. The d_{zx} orbital is stabilized by the interaction with π C and slightly destabilized (poor overlap) by interaction with π O. This HOMO is mainly centered on the metal and corresponds to some backbonding M $\rightarrow$ C. The LUMO is mainly centered on carbon which is electrophilic. The rotation barrier (backbonding) is weak. These orbital schemes depict some limit cases of nucleophilic and electrophilic carbene complex. In practice, a continuum exits between these two limits and the reality is not so clear-cut. As a general rule, Schrock carbenes are derived from high oxidation state metals (Nb(+5), Ta(+5), W(+6)), the M=C double bond is shorter than the M–C single bond by ca. 10 % and the rotation barriers are in the range 10–15 kcal mol^{-1}. The Fischer carbenes are derived from low oxidation state metals (Cr(0), Mo(0), W(0), Fe(0)), the M–C bond is essentially a single bond and the rotation barriers are very low, ca. 0.5 kcal mol^{-1}.

The synthesis of Fischer carbenes supports their theoretical description. The initial synthesis relied on the reaction of organolithium derivatives with metal carbonyls:

for example: L_nM = W(CO)$_5$, Nu$^-$ = RLi and E$^+$ = Me$^+$ (Me$_3$O$^+$)

Since all the chemistry takes place at carbon, the metal must keep its (0) oxidation state. The intermediate acyl anion has a delocalized negative charge between O (hard) and M (soft), thus a hard alkylation reagent is needed like a trialkyloxonium salt. This chemistry can be transposed to thiocarbonyl derivatives as shown in an example:

Acyl derivatives can also be used as shown:

$$L_nM-\overset{\overset{R}{|}}{C}=O \; + \; E^+ \quad \longrightarrow \quad L_nM\overset{+}{=}\overset{R}{\underset{OE}{\diagdown}}$$

Since the carbon of Fischer carbenes is electrophilic, it can be attacked by nucleophiles. Using this feature, it is possible to exchange the carbene heterosubstituents:

$$L_nM=\overset{OR'}{\underset{R}{\diagup}} \; + \; Nu^- \quad \longrightarrow \quad \left[L_nM\overset{-}{-}\overset{OR'}{\underset{R}{\diagdown}}Nu \right] \overset{H^+}{\longrightarrow} \quad L_nM=\overset{Nu}{\underset{R}{\diagup}} \; + \; R'OH$$

$$Nu^- = H^-, \; R^-, \; RS^-, \; R_2N^-$$

It must be noticed that the resulting carbene complex is, in some cases, neither a typical Fischer, nor a typical Schrock carbene complex. These 'hybrid' carbenes have a low stability. The attack of vinyl or alkynyl metal complexes by electrophiles is another route to carbene complexes:

$$M\diagdown_{CH_2} \overset{H^+}{\longrightarrow} \overset{+}{M}=\diagdown_{CH_3}$$

$$M\!\!\!\equiv\!\!\!-R \overset{H^+}{\longrightarrow} \overset{+}{M}\!\!\!\equiv\!\!\!CHR \overset{MeOH}{\longrightarrow} \overset{+}{M}=\overset{OMe}{\underset{CH_2R}{\diagup}}$$

Still another possibility is the electrophilic abstraction of X^-:

$$M-CF_3 \overset{BF_3}{\longrightarrow} \left[\overset{+}{M}=CF_2 \right] \left[BF_4^- \right]$$

$$M = Fe(CO)(PPh_3)Cp$$

In a very limited number of cases, the reaction of a gem-dihalide with a metal dianion can also afford a carbene complex. In the example shown, it is obviously the 2π aromaticity of the cyclopropenylium cation which drives the formation of the carbene complex which is of the Fischer type (electrophilic) even though it has no heteroatom substituent.

$$Na_2Cr(CO)_5 \; + \quad \overset{Ph}{\underset{Ph}{\underset{Cl}{\overset{Cl}{\diagup\!\!\!\diagup}}}} \quad \longrightarrow \quad (OC)_5Cr=\overset{Ph}{\underset{Ph}{\diagup\!\!\!\diagup}}$$

The final solution is to use stable free singlet carbenes. Most of them derive from the imidazole ring and are 6π aromatic systems [3]:

$$\overset{Ar}{\underset{Ar}{\diagup}}: \quad \overset{[M(cod)_2]}{\longrightarrow} \quad \overset{Ar \quad Ar}{\underset{Ar \quad Ar}{\diagup}}=M=\diagdown$$

M = Ni, Pt
cod = cyclooctadiene
Ar = 2,4,6-$Me_3C_6H_2$

The Schrock carbene complexes are less accessible than the Fischer type. Their main synthesis relies on the α-H elimination:

Since the β-H is easier than the α-H elimination, the carbon substituent must be devoid of β-H. R and CH must be cis. The reaction is favored by steric congestion on the metal since it leads to decompression. This is an intramolecular, first-order reaction. The first step involves a 1,2 migration of H from C to M, followed by a reductive elimination of RH, hence M must be electron deficient. The following example is representative.

$$Cl_2CpTa(CH_2CMe_3)_2 \rightarrow Cl_2CpTa=CHCMe_3 + CMe_4$$
$$14e$$

The elimination of CMe_4 is preferred to the elimination of HCl for thermodynamic reasons. The Ta–Cl bond is much stronger than the Ta–C bond.

The best detection tool for carbene complexes is ^{13}C NMR spectroscopy. The Fischer carbenes are found in the range 270–400 ppm (vs. tetramethylsilane), whereas the Schrock carbenes are in the range 225–300 ppm. There is no correlation between the charge at C and the chemical shift.

The thermal decomposition of Fischer carbenes yields a mixture of alkenes.

is not formed

The reaction does not proceed via a free carbene. In the example shown, the free carbene would yield some oxetanone by ring contraction. Mild oxidation (air, Ce(+4), Me_2SO, etc.) yields the corresponding carbonyl derivatives:

Organolithium derivatives can perform the exchange of the carbenic heteroatomic substituent as shown earlier. They also can act as a base and abstract a proton located in α position to the carbenic carbon. This reaction is favored over the substitution at low temperature.

$$(OC)_5Cr = C\begin{smallmatrix} OMe \\ \\ CH_3 \end{smallmatrix} \quad \xrightarrow[\text{THF, -70°C}]{\text{BuLi}} \quad (OC)_5Cr = C\begin{smallmatrix} OMe \\ \\ \ominus \\ CH_2 \end{smallmatrix} \quad \xrightarrow{\text{PhCHO}}$$

$$(OC)_5Cr = C\begin{smallmatrix} OMe \\ \\ CH_2 - CHPh \\ | \\ O^{\ominus} \end{smallmatrix} \quad \xrightarrow[-H_2O]{H^+} \quad (OC)_5Cr = C\begin{smallmatrix} OMe \\ \\ CH = CHPh \end{smallmatrix}$$

The reaction with phosphines can follow different pathways. The initial attack takes place at the electrophilic carbenic carbon as expected:

$$(OC)_5M = C\begin{smallmatrix} X \\ \\ R' \end{smallmatrix} \quad + \; PR_3 \quad \longrightarrow \quad (OC)_5M - \overset{-}{C} \begin{smallmatrix} X \\ | \\ \\ + \\ PR_3 \end{smallmatrix} R'$$

Then, either the phosphine migrates onto the metal or a second equivalent of phosphine displaces the Wittig ylide from the metal:

$$(OC)_5M - \overset{-}{\underset{\overset{+}{PR_3}}{C}} \begin{smallmatrix} X \\ | \\ \\ \end{smallmatrix} R' \quad \xrightarrow{\Delta} \quad CO \; + \; (OC)_4(R_3P)M = C\begin{smallmatrix} X \\ \\ R' \end{smallmatrix}$$

$$\xrightarrow{PR_3} \quad (OC)_5M - PR_3 \; + \; R_3\overset{+}{P} - \overset{-}{C}\begin{smallmatrix} X \\ \\ R' \end{smallmatrix}$$

Wittig ylides can be viewed as a source of nucleophilic carbenes. As such, they can be coupled with electrophilic Fischer carbenes to give alkenes:

$$(OC)_5M = C\begin{smallmatrix} X \\ \\ R' \end{smallmatrix} \quad + \; R_3\overset{+}{P} - \overset{-}{C}HR^2 \quad \longrightarrow \quad (OC)_5M - \overset{-}{\underset{\overset{+}{R_3P - CHR^2}}{C}}\begin{smallmatrix} X \\ | \\ \\ \end{smallmatrix} R'$$

$$\longrightarrow \quad (OC)_5M - PR_3 \quad + \quad \begin{smallmatrix} X \\ \\ R' \end{smallmatrix} C = CHR^2$$

The reaction with alkenes is very sensitive to the substitution schemes of both reagents. In some cases, the alkene reacts as a nucleophile. The attack takes place at the carbenic carbon and the final product is a cyclopropane. The reaction is stoichiometric.

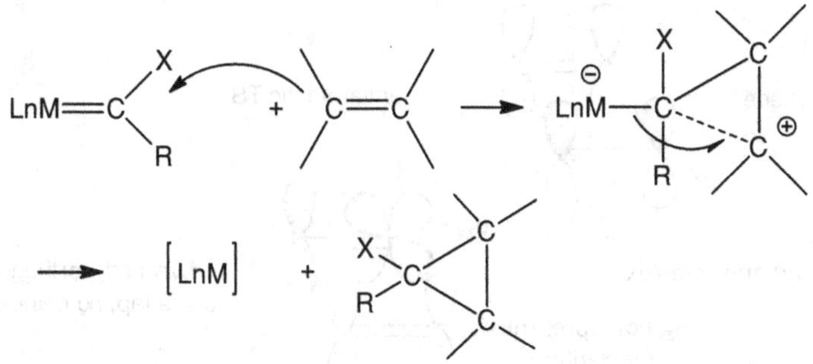

In some other cases, a metathesis is observed, leading to a new carbene complex and a new alkene. The reaction starts by the complexation of the alkene at the metal and is catalytic.

Complexation at the metal

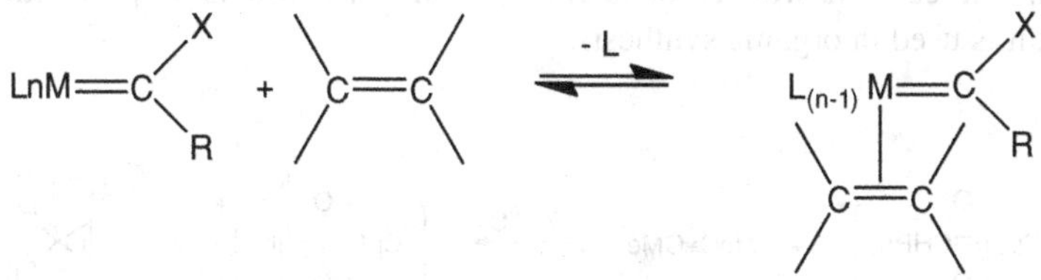

Formation of a metallacyclobutane

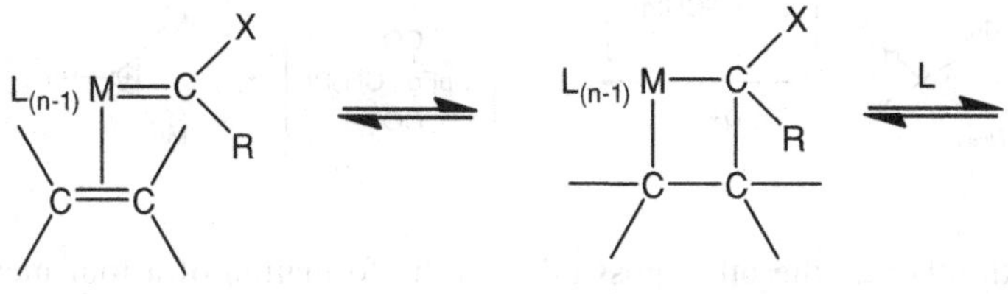

Cycloreversion

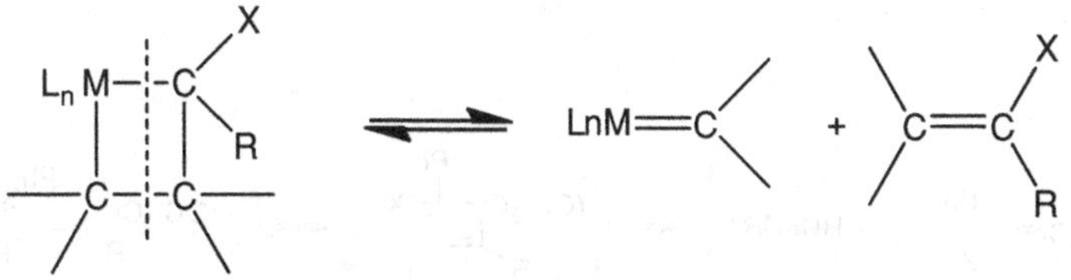

The [2 + 2] concerted cycloadditions are forbidden by the Woodward–Hoffmann rules, so the question which arises is why it becomes possible with a transition metal. The answer is that the interdiction results from the 4π antiaromaticity of the transition state in the alkene + alkene case. The transition metal switches off the delocalization in the transition state which is not antiaromatic anymore. Thus, the [2 + 2] cycloaddition becomes possible.

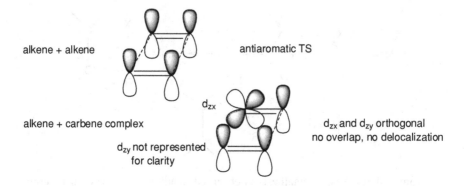

alkene + alkene   antiaromatic TS

alkene + carbene complex d_{zx} and d_{zy} orthogonal
                                                                      no overlap, no delocalization

d_{zy} not represented
for clarity

The reaction with alkynes can also lead to a three-membered ring, especially when the initially formed cyclopropene can aromatize to give a 2π cyclopropenylium ion. The cationic iron–carbene complex shown below is so powerful a reagent that it is used in organic synthesis.

As with alkenes, the other possibility is the formation of a four-membered ring. The metallacyclobutene can evolve by ring opening to give a metallabutadiene. With a phenyl substituent on the carbene, the final product is a naphthol.

The intermediate ketene complex can be trapped by a nucleophile.

$$(OC)_5Cr=\overset{R}{\underset{X}{\diagdown}} \quad \underset{hv}{\rightleftharpoons} \quad \left[O=C=\overset{R}{\underset{\underset{Cr(CO)_4}{X}}{\diagdown}} \right] \quad \overset{NuH}{\longrightarrow} \quad O=C\overset{R}{\underset{\underset{Nu\ H}{}}{}}X$$

With an imino substituent on the carbene, a pyrrole can be obtained in good yield through a formal [3 + 2] cycloaddition.

$$(OC)_5Cr=\overset{N=\diagup Ph}{\underset{Ph}{\diagdown}} \quad \overset{EtOC\equiv CPr}{\longrightarrow} \quad \underset{Ph}{\overset{Pr\quad OEt}{\diagup}}\underset{\underset{H}{N}}{}Ph \quad (95\%)$$

Numerous other synthetic applications of Fischer carbenes are known. Such is not the case for Schrock carbenes which are more difficult to prepare and for which the variety of substituents is more limited. Their widest application is as catalysts for the metathesis of alkenes. This point will be discussed later. Apart from that, they behave as super Wittig ylides. The driving force of the Wittig reaction is the formation of the strong P=O bond (ca. 130 kcal mol^{-1}). With Schrock carbenes, the driving force is much higher: M=O ca. 170–180 kcal mol^{-1}. The examples given hereafter have no equivalents with Wittig ylides. In practice, only the Tebbe's reagent, a stabilized form of [CP$_2$ Ti=CH$_2$], has found some use in organic synthesis.

$$(Me_3C-CH_2)_3Ta=CH-CMe_3$$

$$\xrightarrow{RC(O)OEt} \quad \overset{R}{\underset{EtO}{\diagdown}}C=CHCMe_3$$

$$\xrightarrow{CO_2} \quad Me_3CCH=C=CHCMe_3$$

$$\xrightarrow{HC(O)NMe_2} \quad \overset{H}{\underset{Me_2N}{\diagdown}}C=CHCMe_3$$

$$\xrightarrow[\text{(addition)}]{RC(O)Cl} \quad (Me_3CCH_2)_3Ta\overset{Cl}{\underset{\underset{O=CR}{\diagdown}}{—CHCMe_3}} \quad \rightarrow \quad (Me_3CCH_2)_3Ta\overset{Cl\quad R}{—O—C}=CHCMe_3$$

$$\xrightarrow{RC\equiv N} \quad (Me_3CCH_2)_3Ta\overset{}{\underset{\underset{N=CR}{\diagdown}}{—CHCMe_3}} \quad \rightarrow \quad (Me_3CCH_2)_3Ta=N\overset{R}{—C}=CHCMe_3$$

2.6 Metal Carbynes

Metal carbyne complexes can be obtained from Fischer carbene complexes by electrophilic abstraction:

$$M{=}\mathrm{C}\genfrac{}{}{0pt}{}{X}{R} + E^+ \longrightarrow \overset{+}{M}{\equiv}CR + EX \longrightarrow X{-}M{\equiv}CR$$

$$X = OMe,\ OH,\ Cl...\ E^+ = BX_3,\ Ag^+$$

Example:

$$(OC)_5Cr{=}\genfrac{}{}{0pt}{}{OMe}{Me} + 2\,BCl_3 \longrightarrow (OC)_5\overset{+}{Cr}{\equiv}CMe + BCl_2OMe$$

$$BCl_4^-$$

$$Cl(OC)_4Cr{\equiv}CMe$$

Another, less versatile method, relies on a double α-H elimination promoted by a very high steric congestion on the metal:

$$WCl_6 + 6LiCH_2SiMe_3 \rightarrow (Me_3SiCH_2)_3\,W{\equiv}CSiMe_3 + 2Me_4Si$$

Other possibilities are the metathesis between M≡M and C≡C triple bonds and the deprotonation of some carbene complexes M=CHR.

From a theoretical standpoint, a free carbyne [RC] has three frontier orbitals, one σ orbital on the z axis and two degenerate orthogonal p orbitals p_x and p_y of slightly higher energy. Two main electronic configurations are possible. Each of the three available electrons can occupy a different orbital. This configuration behaves as an X_3 ligand (OS + 3, see the tungsten example before). In the second configuration, two electrons occupy the σ orbital and one a degenerate p orbital. The carbyne behaves as an LX ligand (OS + 1). In practice, the difference between the two types of carbyne complexes is minimal.

The chemistry of carbyne complexes is less developed than the chemistry of carbene complexes. The UV irradiation leads to a transient 1e carbyne complex that can be protonated:

$$M{\equiv}CR \xrightarrow{h\nu} \left[M{-}C\genfrac{}{}{0pt}{}{\cdot\cdot}{\backslash R}\right] \xrightarrow{H^+} \overset{+}{M}{=}C\genfrac{}{}{0pt}{}{H}{R}$$

Cycloaddition onto the M≡C triple bond is observed with sulfur, SO_2 and [PtL$_2$]. Metathesis of alkynes is possible and is now used in organic synthesis. The coupling of two carbynes within a single coordination sphere to give an alkyne has been demonstrated. The deprotonation of a methylidyne complex to give a carbon complex has been achieved.

$$Cl_3(Et_3PO)W{\equiv}C^tBu + PhC{\equiv}CPh \xrightarrow{70\,^\circ C} Cl_3(Et_3PO)W{\equiv}CPh + PhC{\equiv}C^tBu$$

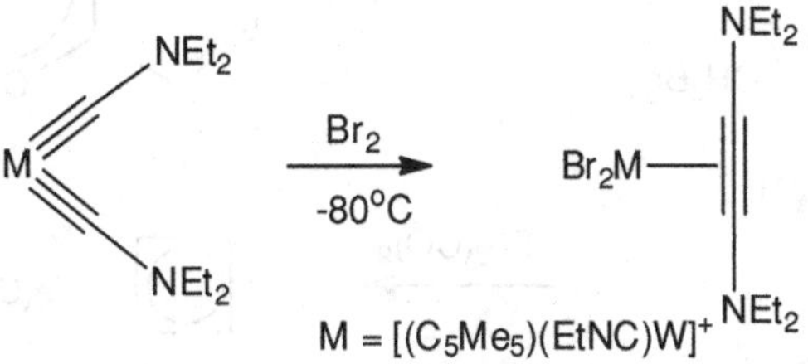

$$M = [(C_5Me_5)(EtNC)W]^+$$

$$(R_2N)_3Mo{\equiv}CH \xrightarrow[THF]{PhCH_2K} [(R_2N)_3Mo{\equiv}C]^-$$

2.7 Some Representative π Complexes

The field of π complexes is so enormous that it is impossible to cover it in any detail within the restricted size of this book. A few representative complexes have been chosen for which some applications in organic chemistry have been developed.

2.7.1 η^4-Diene-Irontricarbonyls

The first complex of this class was discovered in 1930. The preparation is quite simple. It suffices to heat a free diene with an iron carbonyl. The more reactive $[Fe_2(CO)_9]$ is preferred to the rather inert $[Fe(CO)_5]$. It is possible to exchange a complexed diene with a more reactive free diene:

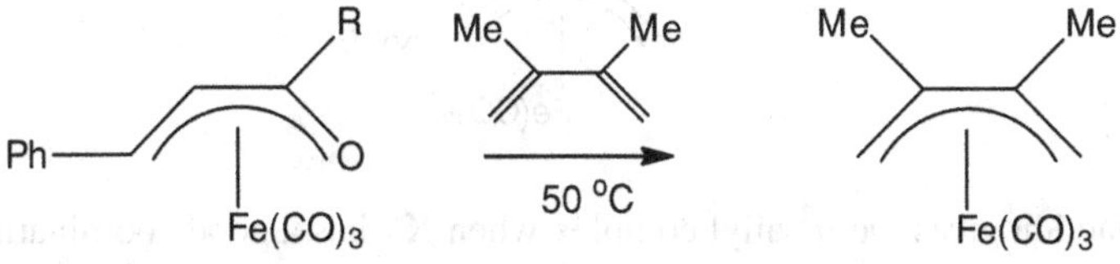

It is also possible to prepare complexes deriving from dienes that are unstable in the free state using *inter alia* the reaction of sodium tetracarbonylferrate with the appropriate dihalides:

Contrary to the antiaromatic and unstable free cyclobutadiene, its irontricarbonyl complex is stable and aromatic: it formally derives from the 6π aromatic cyclobutadiene dianion. Its discovery in 1958 was a highlight of transition metal chemistry. Its decomplexation by mild oxidation (Ce($+4$)) has served to generate free transient cyclobutadiene and develop its chemistry.

In the butadiene complex, the diene is planar and the iron atom is located on an axis perpendicular to this plane at ca. 1.64 Å. The C–C bonds of the diene are equal at ca. 1.40–1.42 Å. The strength of the iron–diene bond is average at ca. 48 kcal mol^{-1}. The NMR data are quite striking. The terminal CH$_2$ protons are found at 0.22 (endo) and 1.90 (exo) ppm and the carbon at 41.6 ppm. This huge shielding effect is quite characteristic of the π complexation.

The most significant chemistry is the reactivity toward electrophiles. Two pathways are possible, leading either to η^3-allyl or to functional η^4-diene complexes.

The reaction gives the η^3-allyl complex when X$^-$ has a good coordinating ability (Cl$^-$) and the diene complex when X$^-$ is not a ligand (AlCl$_4{}^-$). The acetylation (EX=MeC(O) AlCl$_4$) of butadiene-irontricarbonyl is 3,850 times faster than the acetylation of benzene. The formylation (E=CHO) by HC(O)N(Me)Ph + P(O)Cl$_3$ is also possible. Another point of interest is that, since the two faces of the diene are nonequivalent in these complexes, the terminal carbon atoms become chiral when they are substituted. The complex is, indeed, different from its image as shown. Hence, the possibility to develop some applications in enantioselective synthesis has been explored.

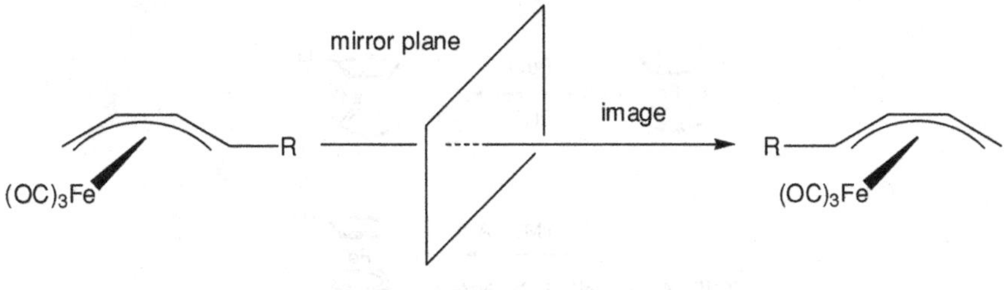

Finally, the easily formed η^5-coordinated α-carbocations can serve to perform electrophilic substitutions on electron-rich arenes.

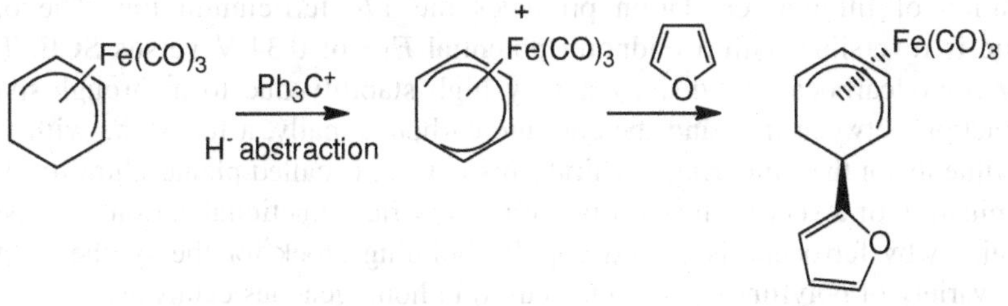

2.7.2 Ferrocene

The discovery of ferrocene in 1951 was a major turning point of transition metal chemistry. This stable orange solid is easily obtained by reacting [CpM] (M = Li, Na, MgBr...) with $FeCl_2$. All the Fe–C bonds are equal at 2.06 Å. The two parallel Cp rings rotate almost freely (barrier ca. 3.8 kJ mol^{-1}). The NMR data shows the characteristic shielding associated with π complexation: δ(H) 4.04, δ(C) 68.2 ppm (in CDCl$_3$). The system is highly aromatic. Both electrophilic acylation and formylation are possible. In fact, the acetylation by MeC(O)Cl + Al Cl$_3$ is 3×10^6 times faster than that of benzene.

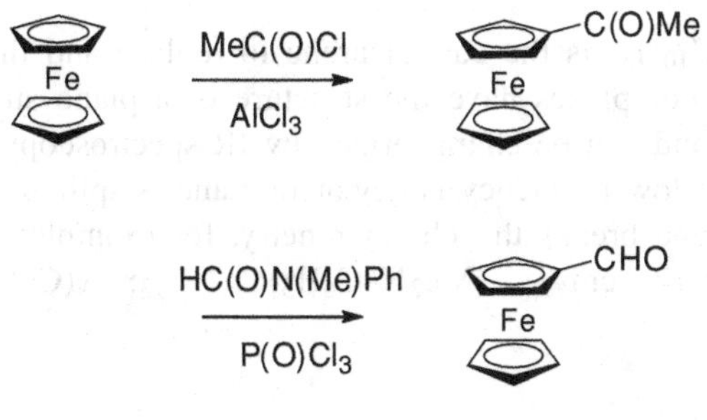

The 1,1′-diacylation on the two rings is also possible but the 1,1′-diformylation is not. The protonation with strong acids takes place at iron as shown by NMR: δ(H)—12 ppm. Mono- and dilithiations are feasible:

$$Fe \xrightarrow[\text{Et}_2\text{O}]{\text{BuLi}} Fe\text{—Li}$$

$$\xrightarrow{\text{TMEDA}} Fe \begin{array}{l}\text{—Li}\\\text{—Li}\end{array}$$

$$\text{TMEDA} = \text{Me}_2\text{NCH}_2\text{CH}_2\text{NMe}_2$$

Aminomethyl substituents direct the metalation toward the α position by chelation of lithium. Oxidation produces the 17e ferricinium ion. The oxidation is reversible with a midpoint potential $E_{1/2}$ of 0.34 V versus SCE. The α-ferrocenylcarbocations display a very high stability due to a through-space interaction between iron and the cationic carbon. Finally, a ferrocene with two substituents on the same ring is chiral; this is what is called planar chirality. The combination of a very high stability with a very rich functionalization chemistry explains why ferrocene is such a popular building block for the synthesis of a wide variety of polyfunctional ligands used in homogeneous catalysis.

2.7.3 η^6-Arene-Chromiumtricarbonyls

These species are easily obtained by allowing a free arene to react with chromium hexacarbonyl. The reaction is favored by donor substituents and inhibited by acceptor substituents. In difficult cases, $Cr(CO)_6$ is replaced by $Cr(CO)_3L_3$ ($L = CH_3CN$, for example). The decreasing stability order is:

$$Me_6C_6 > Me_3C_6H_3 > Me_2N-C_6H_5 > Me_2C_6H_4 > C_6H_6 > MeC\,(O)\,C_6H_5$$
$$> XC_6H_5\,(X = Cl,\ F) > C_{10}H_8$$

Naphthalene ($C_{10}H_8$) is the easiest arene to replace and this has synthetic applications. The complexes have the structure of a piano stool (Cr–benzene distance 1.73 Å) and can be characterized by IR spectroscopy. They give two bands, the band at low frequency is degenerate and is split when a bulky substituent on the arene breaks the C_{3v} symmetry, for example: $(C_6H_6)Cr(CO)_3$: ν(CO) 1987, 1917 cm^{-1}; $(Me_2N-C_6H_5)Cr(CO)_3$: ν(CO) 1969, 1895, 1889 cm^{-1}.

Since the Cr(CO)$_3$ group is strongly electron withdrawing, it activates the arene and its substituents toward the attack by nucleophiles.

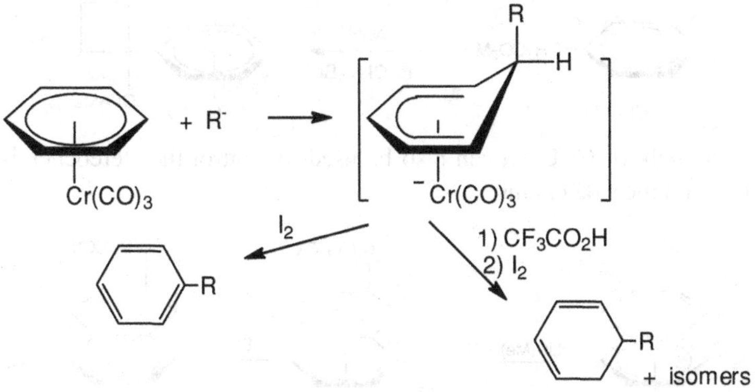

Decomplexation is achieved by mild oxidation. This scheme has numerous synthetic applications. Some of them are shown here.

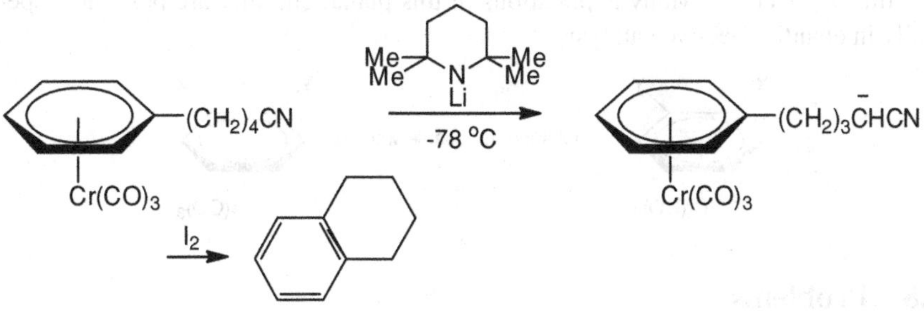

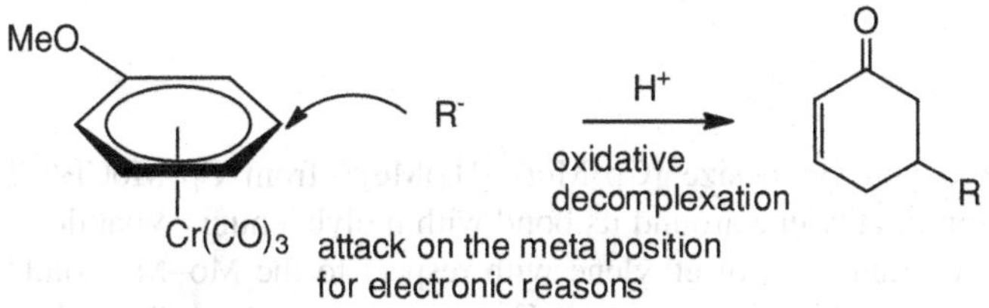

attack on the meta position
for electronic reasons

Lithiation needs a directing group in order to be regioselective.

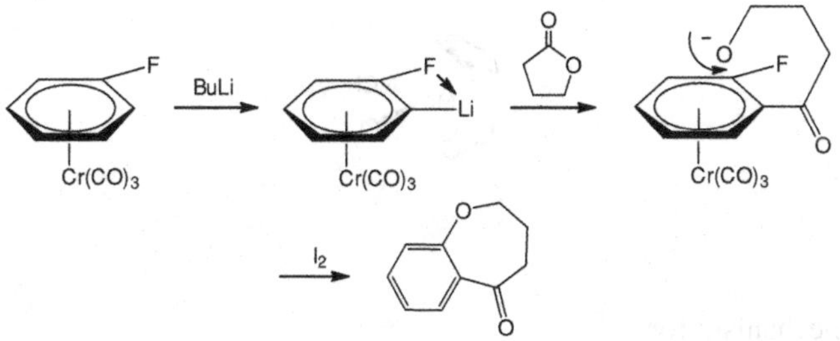

The $Cr(CO)_3$ complexing group also activates the benzylic positions, allowing a facile dimetalation:

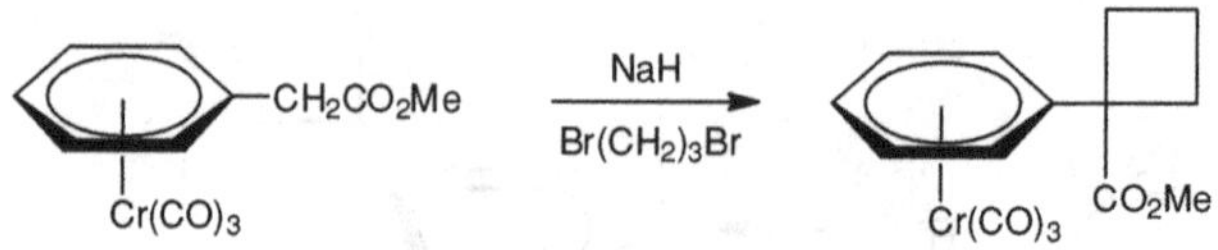

The steric bulk of $Cr(CO)_3$ can also be used to control the stereochemistry of the reactions on the side chains:

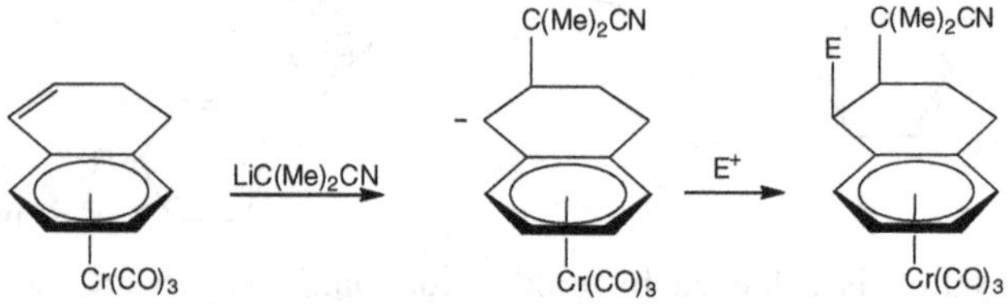

As in the case of ferrocene, an *ortho*-disubstituted derivative with two different substituents is chiral. Many applications of this planar chirality are possible, especially in enantioselective catalysis.

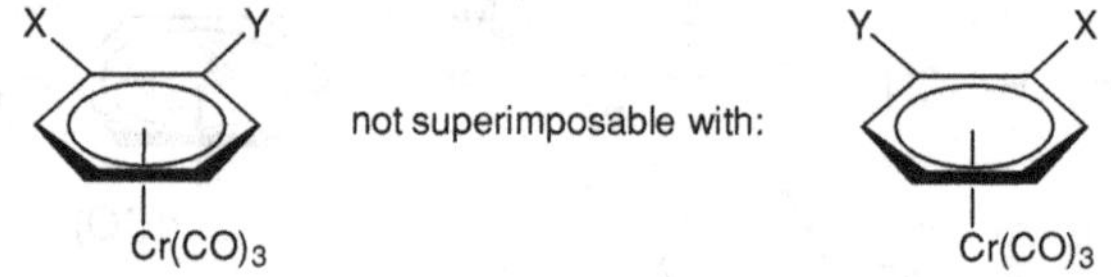

not superimposable with:

2.8   Problems

II.1

How would you synthesize $[Cp_2Mo(C_2H_4)Me]^+$ from Cp_2MoCl_2? There is no free rotation of ethylene around its bond with molybdenum. What does that mean? What is the orientation of ethylene with respect to the Mo–Me bond? (use arguments based on orbital interactions). How can you establish this orientation using a classical spectroscopic technique. You have to recall that Cp_2MoCl_2 has a bent metallocene structure.

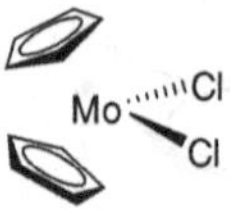

II.2

Propose a mechanism for:

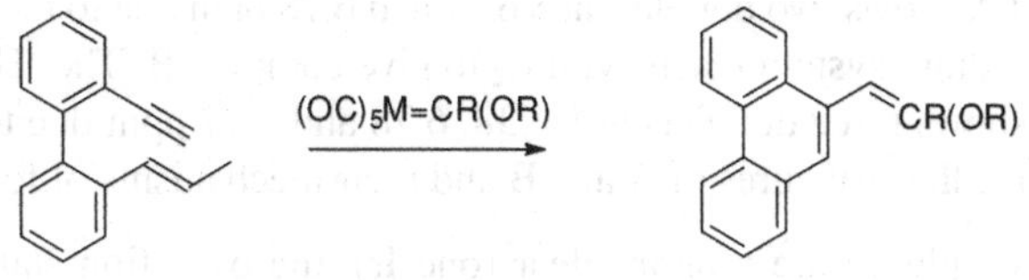

II.3

The following stoichiometric chemistry has been performed by Barluenga:

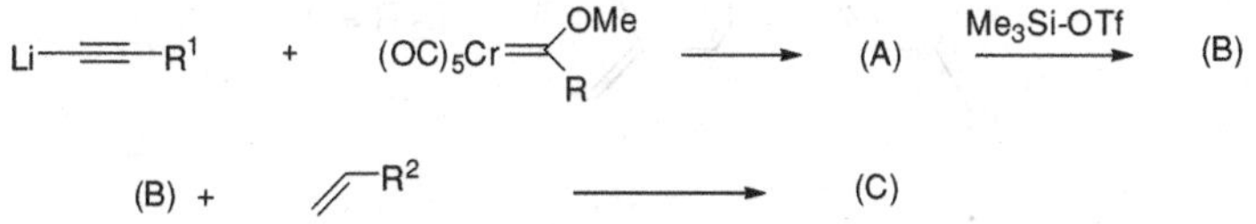

What are the formulas of the chromium complexes (A) and (B) (B does not contain silicon). What are the formula and the two possible stereochemistries of the cyclic organic product (C)

II.4

The reaction of the cationic carbyne complex (A) with a secondary phosphine in the presence of an amine gives the ketene complex (D) (Lugan, Organometallics 2011):

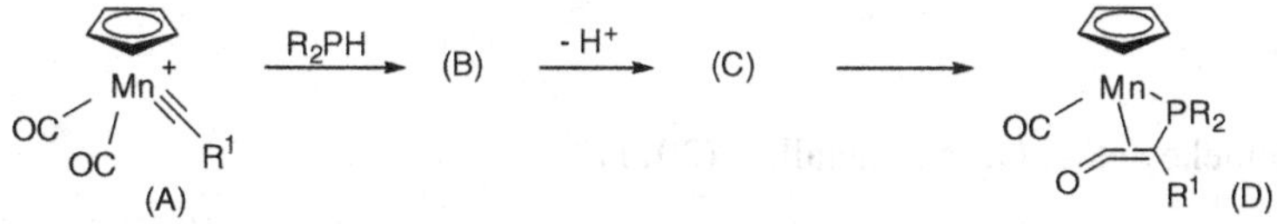

(1) Give the formula of the intermediates (B) and (C).
(2) Propose a mechanism for the last step. What analogy do you find in chromium Fischer carbene chemistry?

II.5

Cp* stands for η^5-C$_5$Me$_5$.

(a) Propose a structure for the complex [IrCl$_2$Cp*]$_2$. Discuss the electron count.
(b) Upon reaction with NaOAc (OAc = acetate), [IrCl$_2$Cp*]$_2$ gives a monomeric complex. Propose a structure for this monomer.

(c) [IrCl$_2$Cp*]$_2$ + NaOAc reacts with L$_1$ to give a single complex A. The ^{1}H NMR
     spectrum of A shows two doublets at δ 6.41 and 6.78 ppm due to the pyrrole ring.
(d) The same iridium system reacts with L$_2$ to give complex B. The ^{1}H NMR spec-
     trum of B shows three doublets at δ 6.38, 6.78 and 7.17 ppm due to the pyrrole
     protons. Give the structures of A and B and their mechanisms of formation.

Note: The complexes are mononuclear (one Ir), the oxidation state of Ir stays
identical, the backbone of L is preserved.

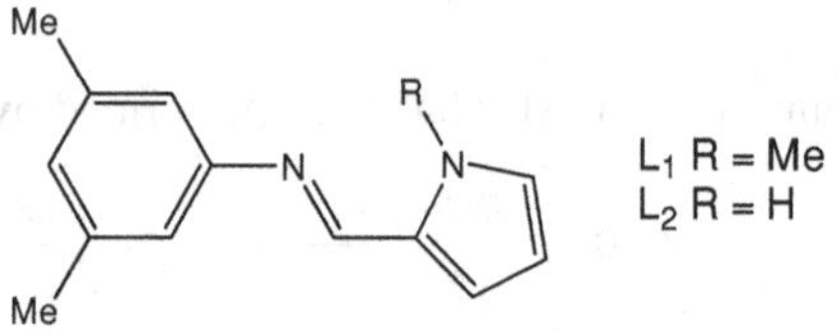

L$_1$ R = Me
L$_2$ R = H

II.6

(1) A ruthenium catalyst represented by [Ru] gives two complexes with a primary
     alkyne R–CC–H. One is a carbene complex. Give the formula of these two
     complexes and explain the formation of the carbene complex.
(2) The reaction of these two complexes with the carboxylate anion R^{1}CO$_2$$^-$
     yields three products after protonation and loss of [Ru]:

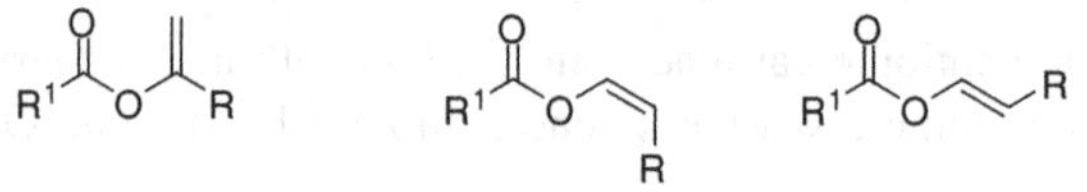

Explain the formation of these products

II.7 Heinekey et al. Organometallics (2011)

The reaction of the Ir(I) complex (A) with (B) gives a square pyramidal Ir(III)
complex. What are the formula and the number of electrons of (C)?

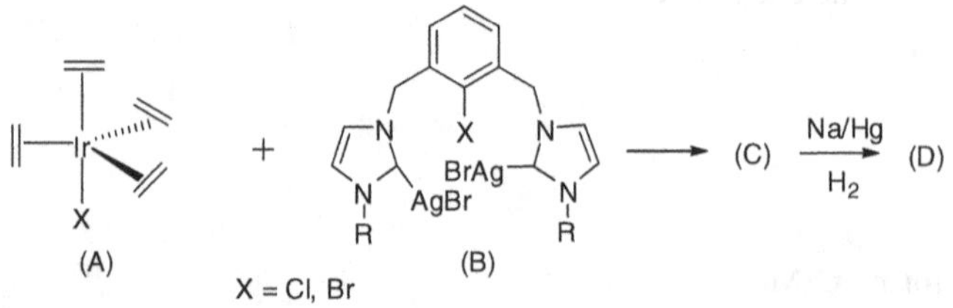

X = Cl, Br

Upon reaction with sodium under hydrogen, (C) gives the octahedral Ir(III)
complex (D). This complex shows a single ^{1}H resonance at −9 ppm (4H). What
is the formula and the number of electrons for (D)? Why is only one resonance

observed for the hydride? The reaction of (D) with PMe_3 gives another complex (E) with two hydride resonances (1H + 1H), whereas the reaction with pyridine gives (F) with a single hydride resonance (2H). Why?

II.8

The synthesis of an organometallic pyrylium salt has been described (Bruce et al. Organometallics 2011):

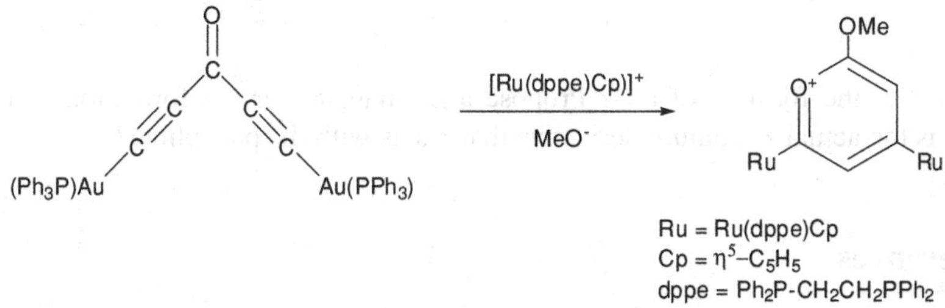

Ru = Ru(dppe)Cp
Cp = η^5-C_5H_5
dppe = Ph_2P-$CH_2CH_2PPh_2$

The proposed mechanism involves first a metal exchange to give (A), then a double protonation to give a biscarbene complex (B). The attack of this biscarbene by methylate anion gives a four-membered OC_3 ring (C) whose rearrangement gives the pyrylium salt. Give the structures of (A), (B), and (C) and comment on the various steps. What is the rearrangement that transforms (C) into the final product?

II.9

The following reaction sequence has been described (Alcock et al. Organometallics 1991, 10, 231):

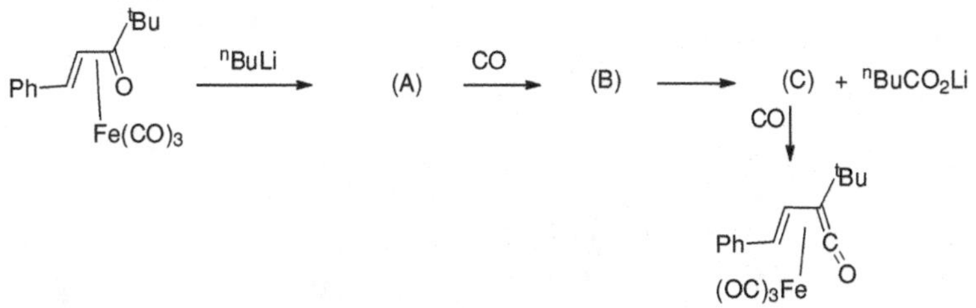

(A), (B), and (C) are Fe–carbene complexes. Propose a formula for (A), (B) and (C) and discuss the mechanism for their formation. What reaction described in the book is reminiscent of this transformation?

II.10 Meunier, Majoral et al. Angewandte (1997)

The following reactions have been performed:

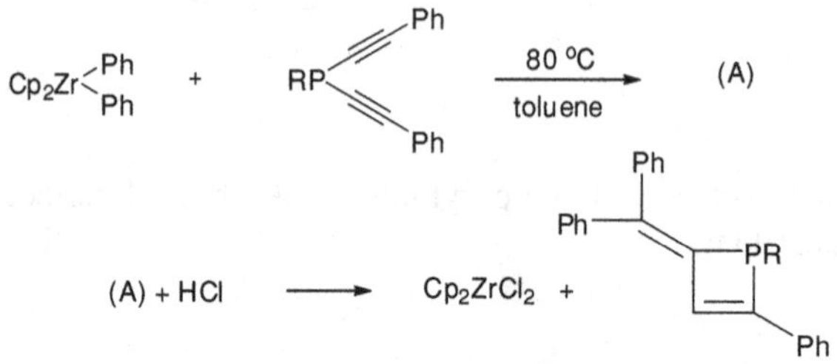

$Cp_2Zr\langle^{Ph}_{Ph}$ + RP (diyne with Ph, Ph substituents) $\xrightarrow[\text{toluene}]{80\ ^{\circ}C}$ (A)

$(A) + HCl \longrightarrow Cp_2ZrCl_2 +$ (four-membered ring product with Ph, Ph, PR, Ph substituents)

What is the formula of (A)? Propose a mechanism for the formation of (A). What is the actual zirconium derivative that reacts with the phosphine?

References

1. Jessop PG, Morris RH (1992) Reactions of transition metal dihydrogen complexes. Coord Chem Rev 121:155
2. Elian M, Hoffmann R (1058) Bonding capabilities of transition metal carbonyl fragments. Inorg Chem 1975:14
3. Crudden CM, Allen DP (2004) Stability and reactivity of N-heterocyclic carbene complexes. Coord Chem Rev 248:2247

Chapter 3
Homogeneous Catalysis

Abstract This chapter can be used as an introduction to homogeneous catalysis. The first section deals with the hydrogenation of alkenes. The basic mechanism is discussed in the case of the Wilkinson catalyst. Then, the asymmetric hydrogenation of functional alkenes is exemplified with chiral ligands such as DIOP, DIPAMP, and BINAP. This section concludes by introducing the phosphino-oxazolines as efficient ligands for the difficult enantioselective hydrogenation of non-functional alkenes. The following two sections describe the hydrosilylation of alkenes and alkynes and the hydrocyanation of alkenes. The important synthesis of adiponitrile (precursor of nylon 6,6) by hydrocyanation of butadiene is more thoroughly discussed. The next section introduces the various processes in use for the hydroformylation of alkenes: Roelen (cobalt), Shell (cobalt + phosphine), Union Carbide (rhodium), and Rhone-Poulenc (rhodium biphasic). Then, the immensely important Ziegler–Natta polymerization of ethylene and propylene is presented. The control of the tacticity of polypropylene by the structure of the zirconocene catalysts is explained. The section on the metathesis of alkenes discusses its various useful versions [ring closing metathesis (RCM) and ring opening metathesis polymerization (ROMP)], introduces the Grubbs catalysts and is completed by the description of the metathesis of alkynes. The section on palladium starts by the uses of Pd(II) in the Wacker process (oxidation of ethylene into acetaldehyde) and in the hydroamination of alkenes. Then, the uses of Pd(0) catalysts in the various C–C cross-coupling reactions (Heck, Stille, Suzuki) are discussed. The section on gold is illustrated by several catalytic transformations of alkynes.

Keywords Hydrogenation • Asymmetric catalysis • Hydroformylation • Polymerization • Metathesis • C–C cross coupling

3.1 Catalytic Hydrogenation

Transition metals have several, easily reached oxidation and coordination states. These characteristics are ideal for the construction of catalysts that are supposed to activate bonds through oxidative addition, create bonds by intramolecular coupling, and extrude products by reductive elimination. In fact, these metals play a

F. Mathey, *Transition Metal Organometallic Chemistry*,
SpringerBriefs in Molecular Science, DOI: 10.1007/978-981-4451-09-3_3,
© The Author(s) 2013

central role in catalysis which is, by far, their main use in organic synthesis. When compared to heterogeneous catalysis where the catalyst is an inorganic solid interacting with the reagents by a variety of sites on its surface, homogeneous catalysis, where, in most cases, the catalyst is a single metal molecule dissolved in the solvent of the reaction, is characterized by the mildness of the experimental conditions and the selectivity and reproducibility of the results. But it has several drawbacks: the catalyst is relatively unstable, difficult to recycle, and very sensitive to poisoning. These questions have to be addressed before any application can be envisaged. Let us recall that the catalyst increases the reaction rate by reducing the energy gap between the starting products and the transition states but cannot initiate a reaction that is not thermodynamically permitted. A typical catalytic cycle is as follows:

$$A + B \quad\xrightarrow{\ K\ }\quad AB$$

$$
\begin{array}{ccc}
 & & B \\
A \longrightarrow KA & & \Big| \\
 & & \downarrow \\
\text{catalyst precursor} \longrightarrow K & & BKA \\
 & & \Big| \text{ A-B coupling} \\
AB \longleftarrow KAB \longleftarrow
\end{array}
$$

Since the real catalyst must coordinate both to A and B, it must have two vacant sites on the metal and a maximum of 14e. Hence, its stability is low. It is generated in situ from a more stable precursor and, since its reactivity is quite high, its concentration in the reaction medium is quite low.

The concerted addition of H_2 onto alkenes is forbidden by orbital symmetries. The LUMO of the alkene is antisymmetric whereas the HOMO of H_2 is symmetric. The role of the catalyst is to circumvent this interdiction. A general simplified catalytic mechanism is shown here.

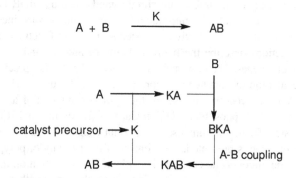

The only irreversible step in this mechanism is the reductive elimination leading to the alkane product. The first classical hydrogenation catalyst was the so-called Wilkinson catalyst [RhCl(PPh$_3$)$_3$]. This stable 16e complex is, in fact, the precatalyst. One triphenylphosphine must be lost to give the genuine 14e catalyst [RhCl(PPh$_3$)$_2$] which is quite reactive and unstable. If not stabilized, it gives the 16e chlorine-bridged dimer which is catalytically inactive. In order to avoid this dead-end, it is necessary to use a weakly coordinating solvent.

$$\begin{array}{ccccc}
Ph_3P & & Cl & & PPh_3 \\
 & \diagdown & | & \diagup & \\
 & Rh & & Rh & \\
 & \diagup & | & \diagdown & \\
Ph_3P & & Cl & & PPh_3
\end{array}$$

Since Rh(+1) is soft according to the Pearson's HSAB theory, the solvents of choice are hard oxygen donors such as MeOH or THF. These solvents (S) give a loose 16e complex [RhCl(PPh$_3$)$_2$(S)] in which (S) is easily displaced by the soft alkene. If triphenylphosphine is replaced by a more basic phosphine such as Et$_3$P, the resulting complex is also inactive because the Rh–P bond becomes too strong to dissociate under the mild catalytic conditions (typically room temperature or slightly above). The coordinating ability of a phosphine is linked to its electron-donating power, roughly measured by its basicity, and to its steric bulk, as measured by the so-called Tolman cone angle.

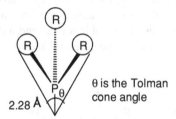

The apex of the cone is on the phosphine axis at 2.28 Å from P (the average Ni–P distance in nickel–phosphine complexes). The cone encircles the three R substituents. When phosphorus has three different substituents, the average angle of the three symmetrical cones is taken. These cone angles measure the steric demand of a phosphine. When the value of the cone angle is high, the M–P distance is long and the strength of the M–P bond is low. The value of these angles (in deg.) varies in a wide range: PH$_3$ 87, (MeO)$_3$P 107, Me$_3$P 118, Et$_3$P 132, Ph$_3$P 145, ^{t}Bu$_3$P 182. Thus, PH$_3$ is a good ligand in spite of its low basicity.

The steric bulk of the alkene also plays a crucial role in the rate of hydrogenation. The order is the following:

Some alkene hydrogenation catalysts can also be used for the reduction of ketones, aldehydes, and imines. Finally, if a hydrogenation catalyst is stable enough, it can also act as a dehydrogenation catalyst at sufficiently high temperature. This is the case for the pincer complex shown below which is able to dehydrogenate alkanes above 200° in the presence of a hydrogen acceptor, although the turn-over numbers (T.O.N = number of catalytic cycles until the catalyst dies) are too low for practical applications.

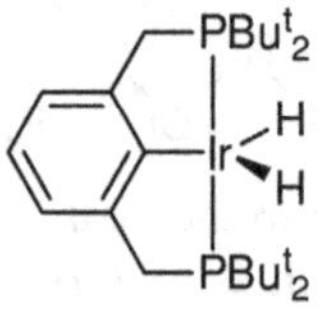

3.2 Asymmetric Hydrogenation [1]

If an sp^2 carbon $=CR^1R^2$ carries two different substituents, the addition of hydrogen will create a chiral center $-CHR^1R^2$. The sp^2 carbon is called prochiral. Asymmetric hydrogenation of prochiral alkenes has become a major tool for creating optically active compounds. The problem is to find a catalyst which is able to distinguish between the two faces of the prochiral alkene:

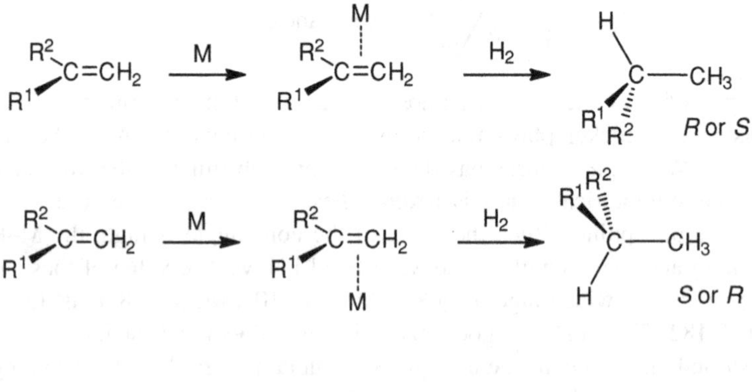

The trick is to introduce an optically active ligand L* in the catalyst ML*. When the catalyst coordinates to the prochiral alkene, the prochiral carbon becomes chiral since it has four substituents. The π complex having two chiral centers (C and L*) exists as a mixture of two diastereomers (R_C, S_L) and (S_C, S_L) if L* is S. These two diastereomers are in fast equilibrium since the π complexation is known to be reversible. These two diastereomers react at different rates with hydrogen, hence the basic principle of asymmetric hydrogenation:

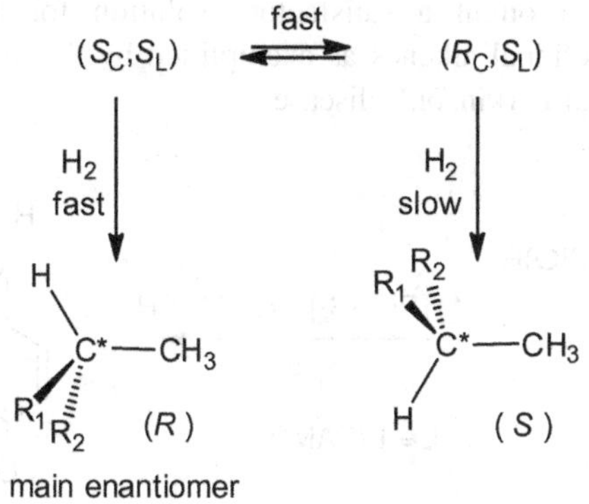

main enantiomer

It must be noticed that C* is S when coordinated to M and R when bonded to H because M has the priority 1 (it has the heaviest atom) and H the priority 4 (it is the lightest atom). If q_R molecules of (R) and q_S molecules of (S) alkane are obtained, the rotation α of the plane of the polarized light is proportional to $q_R - q_S$ and the performance of the enantioselective catalyst is measured by the enantiomeric excess (e.e.):

$$\text{e. e.} = \frac{q_R - q_S}{q_R + q_S} = \frac{\alpha \text{ observed}}{\alpha \text{ max}}$$

Historically speaking, the most important ligands for the enantioselective hydrogenation of alkenes are chelating diphosphines used as their $[L_2RhS_2]^+$ complexes (L_2 = diphosphine, S = solvent). Three of them are especially famous. The first one is the so-called DIOP prepared by Kagan. It derives from the cheap enantiopure tartaric acid. The chiral centers are on the carbon backbone. The second is the DIPAMP of Knowles. Here, the chiral centers are the two pyramidal phosphorus atoms. The third is the BINAP of Noyori. Here, there is no classical chiral center. The chiral information derives from the blocked rotation around the C–C bond bridging the two naphthalene units. This rotation is blocked by the PPh$_2$ groups and the two hydrogen atoms facing each other in the binaphthalene unit. This leads to the existence of two non-superimposable rotamers called atropisomers.

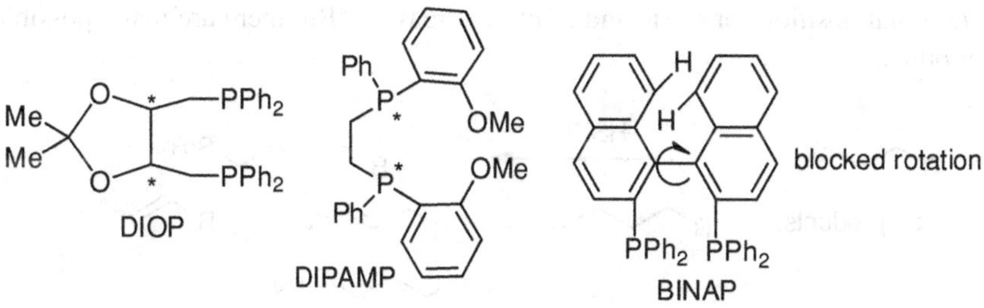

These phosphines brought a satisfactory solution for the enantioselective hydrogenation of functional alkenes as exemplified by the synthesis of L-DOPA used in the treatment of Parkinson's disease.

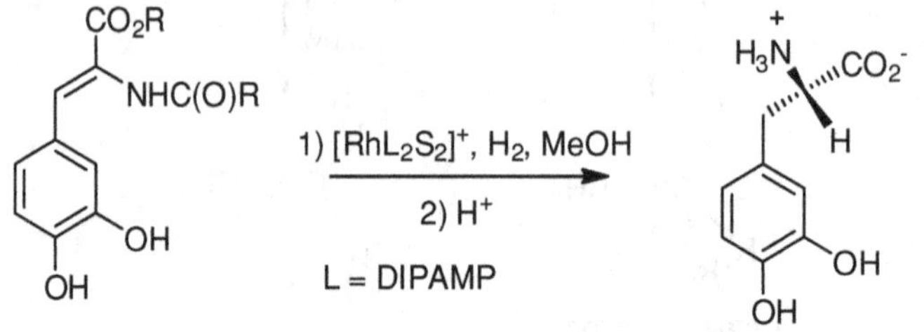

These functional alkenes chelate the catalytic center. This chelation rigidifies the coordination sphere of the metal and favors high e.e.'s. The solution for non-functional alkenes came only around the year 2000 with the use of phosphine derivatives of oxazolines (Pfaltz):

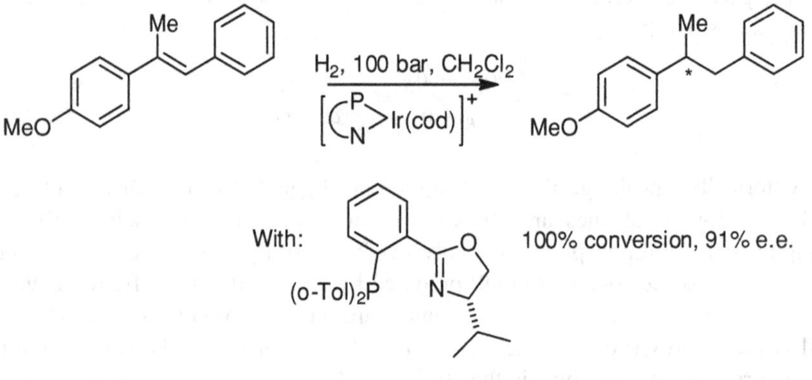

3.3 Hydrosilylation, Hydrocyanation

Alkene hydrosilylation tends to produce the linear silane where silicon occupies the terminal position for steric and electronic reasons. But there are many possible by-products:

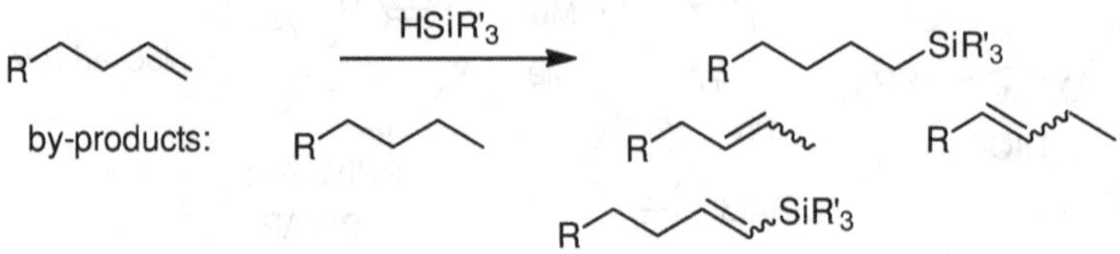

The formation of these products is easy to explain (hydrogenation and isomerisation of the starting alkene) with the exception of the unsaturated silane. Most of the catalysts use platinum as the metal, either as H_2PtCl_6 or as Pt–carbene or Pt–phosphine complexes. The proposed mechanism is depicted hereafter. The H migration explains the formation of the normal hydrosilylation product, whereas the Si migration explains the formation of the silylated alkene.

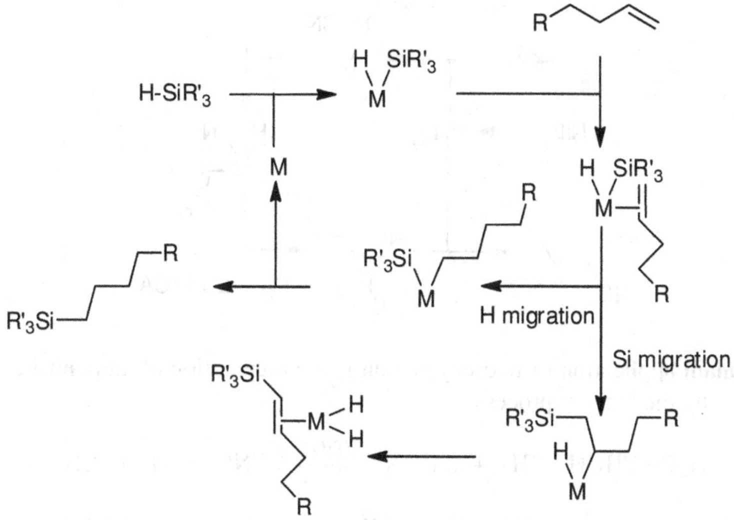

The hydrosilylation of terminal alkynes can produce both the E and Z alkenes. The reason lies in the structure of the intermediate alkenyl–metal hydride:

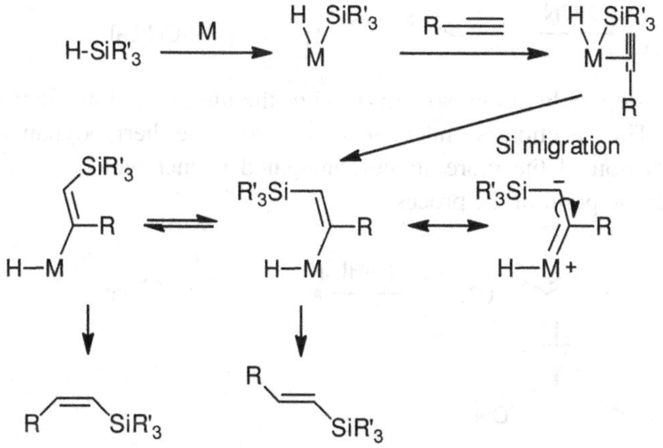

The strong carbene character of this product explains why the barrier to the rotation around its C=C double bond is so low that the Z to E interconversion becomes possible.

In terms of mechanism, hydrocyanation is also related to hydrogenation:

$$RCH{=}CH_2 + H{-}CN \rightarrow RCH_2{-}CH_2{-}CN$$

Here, the catalyst of choice derives from nickel and the ligands are typically triaryl phosphites. The mechanism is shown here.

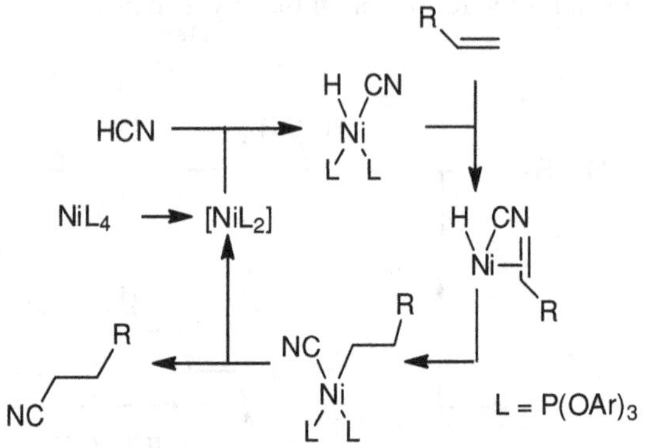

The main application of hydrocyanation is the preparation of adiponitrile from butadiene by the Dupont process.

$$H_2C{=}CHCH{=}CH_2 + 2\ HCN \xrightarrow{[Ni(P(OAr)_3)_4]} NC{-}(CH_2)_4{-}CN$$

The first step involves a 1,4 addition of HCN through an η^3-allylnickel complex:

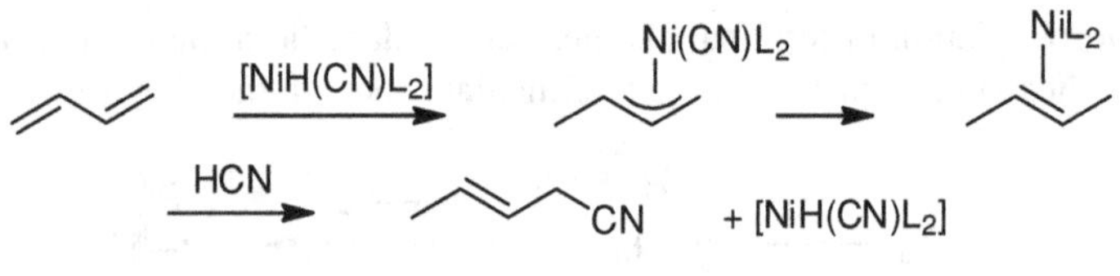

The second step includes an isomerisation of the unsaturated nitrile into its terminal isomer. This reaction is under kinetic control. The thermodynamics would favor the formation of the more stable conjugated isomer which is a dead-end. This is the delicate point of the process.

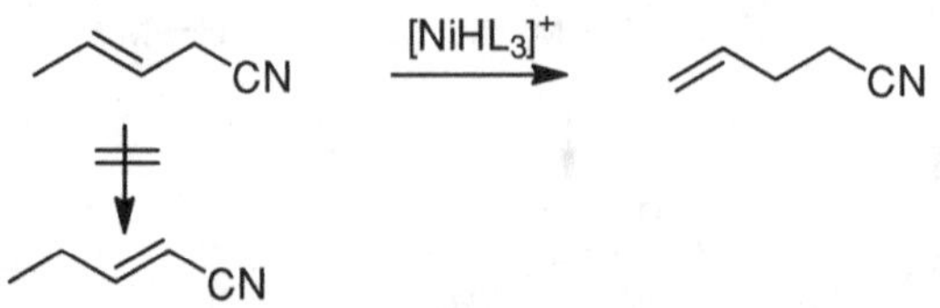

The last step is the classical hydrocyanation of the terminal unsaturated nitrile. The adiponitrile thus obtained is hydrogenated to give the diamine $H_2N(CH_2)_6NH_2$. Another part is hydrolyzed to give the diacid $HO_2C(CH_2)_4CO_2H$. The diamine and the diacid are condensed to give the well-known nylon 6,6.

3.4 Alkene Hydroformylation [2]

This reaction (also called the "oxo" reaction) transforms an alkene into an aldehyde through the addition of $H_2 + CO$:

$$RCH=CH_2 + H_2 + CO \rightarrow RCH_2-CH_2-CHO + RCH(CH_3)-CHO$$
$$\text{(major)} \qquad\qquad \text{(minor)}$$

This transformation is one of the major processes of the chemical industry and the current production of aldehydes through this reaction is close to 7 million tons per year. The reaction was discovered by Roelen in 1938. In the original process, the catalyst was $[HCo(CO)_4]$ as produced in situ from cobalt salts and $H_2 + CO$. The typical reaction conditions are 120–170° and 200–300 bar of $H_2 + CO$. The linearity (ratio of linear to branched aldehydes) is relatively poor at 3–4: 1. The accepted mechanism is shown here. All the steps are reversible. The migration of H from Co to the internal sp^2 carbon of the alkene is controlled by the hydridic nature of H. The steric bulk of cobalt which goes to the terminal position of the alkene is another positive factor. The addition of H_2 leading to the hydrogenolysis of the Co–acyl bond is the rate determining step.

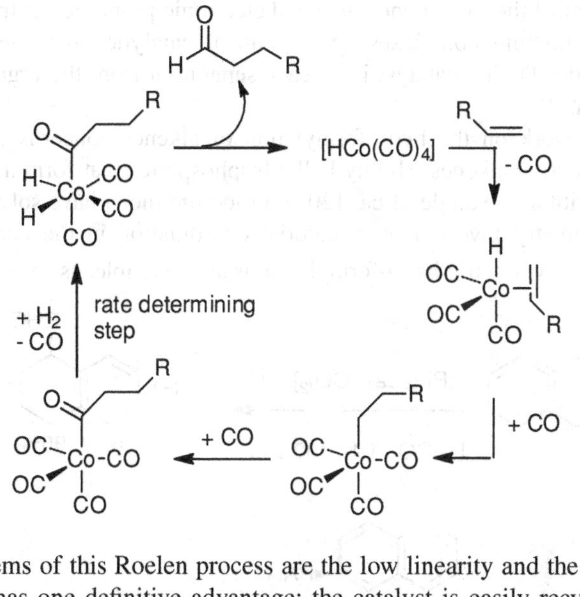

The problems of this Roelen process are the low linearity and the harsh conditions. But it has one definitive advantage: the catalyst is easily recycled through the reaction of the cobalt residues with $H_2 + CO$ followed by distillation of $[HCo(CO)_4]$ (b.p. 47 °C). The Roelen process was modified by Shell in 1966. The cobalt catalyst was improved using the addition of a trialkylphosphine. The new precatalyst $[HCo(PR_3)(CO)_3]$ is more stable, its hydride character is stronger, and its steric bulk is higher. As a result, the experimental conditions are milder (50–100 bar of $H_2 + CO$) and the linearity is better (8: 1). The resulting aldehyde is immediately reduced to the corresponding alcohol. However, two drawbacks have blocked the development of this process, the higher proportion of alkane by-product and the difficult recycling of the catalyst.

The third process was introduced by Union Carbide in 1976. Here the catalyst is [HRh(CO)(PPh$_3$)$_2$]. The new typical conditions are quite mild (H$_2$ + CO 50 bar, temperature 100 °C). The linearity is very high (20: 1) when the reaction is run at room temperature and room pressure, but decreases under the industrial conditions which are needed for a high rate. In practice, a huge excess of triphenylphosphine is used to prevent the decomposition of the catalyst. The details concerning the recovery of the catalyst are not published.

The fourth process was introduced by Rhone-Poulenc in 1981. It works in the aqueous phase. It uses the water-soluble TPPTS instead of triphenylphosphine.

Triphenylphosphine
tris(sulfonate)TPPTS

The process works only with alkenes having enough water solubility (C$_2$–C$_4$). TPPTS has almost the same cone angle and electronic properties as triphenylphosphine. Their rhodium complexes have identical catalytic properties. The huge advantage of the TPPTS catalyst is its easy separation from the organic products of the reaction.

Additional work on the hydroformylation of alkenes concerns the hydroformylation of internal alkenes. Highly bulky bisphosphites that form a 9-membered chelate ring with a bite angle of ca. 120° on rhodium metal have solved this problem. This geometry favors a bis-equatorial coordination in the pentacoordinate rhodium TBP. Asymmetric hydroformylation is also possible, as shown:

[Rh(acac)(CO)$_2$] + L*

H$_2$/CO 1:1 , 100 atm

(–)

CHO

(e. e. 96%)

L* = R,S - BINAPHOS

3.5 Alkene Polymerization [3]

The current production of alkene polymers (mainly polyethylene and polypropylene) is around 100 million tons per year. Polyethylene was discovered as early as 1898 but the first industrial process was only devised in 1935 by Imperial Chemical Industries (ICI) and used high temperatures (ca. 200 °C) and very high pressures (1000–3000 bar). The production started to grow with the discovery by Ziegler in 1953 of titanium catalysts able to polymerize ethylene at room temperature and pressure. A polymerization process always involves the three following steps:

(1) Initiation

$$LnM^+ \ + \ R^- \ (H^-) \ \longrightarrow \ \boxed{LnM-R} \ \left(LnM-H\right)$$

(2) Propagation

$$LnM-R \ + \ \underset{}{\overset{}{>}}C{=}C\underset{}{\overset{}{<}} \ \xrightarrow{K_P} \ \underset{[\,LnM-R\,]}{\overset{}{>}}C{=}C\underset{}{\overset{}{<}} \ \longrightarrow \ LnM-\overset{|}{\underset{|}{C}}-\overset{|}{\underset{|}{C}}-R$$

(3) Termination

$$LnM-\left(\overset{|}{\underset{|}{C}}-\overset{|}{\underset{|}{C}}\right)_n-R \ \xrightarrow{K_T} \ \left(\overset{|}{\underset{|}{C}}-\overset{|}{\underset{|}{C}}\right)_n \ + \ \boxed{LnM-R}$$

The length of the polymer depends on the ratio K_P/K_T. The higher the ratio, the longer the polymer. The ease of polymerization depends on the coordination ability of the alkene. The smaller, the better. The order is: ethylene > propene > 1-butene. There are several termination mechanisms. The major one is the already discussed β-H elimination:

$$LnM\underset{CH_2}{\overset{H\overset{\displaystyle H}{\diagdown}}{\diagup}}C-\left(H_2C-CH_2\right)_n-R$$

$$\longrightarrow \ LnM-H \ + \ H_2C{=}CH-\left(CH_2\right)_m-R$$

Two other reactions can also intervene, the hydrogenolysis of the M–C bond (it is impossible to prevent the formation of hydrogen) and the intramolecular H transfer:

$$LnM—\left(H_3C—CH_2\right)_n—R \ + \ H_2$$

$$\longrightarrow \quad LnM—H \ + \ H—\left(CH_2\right)_m—R$$

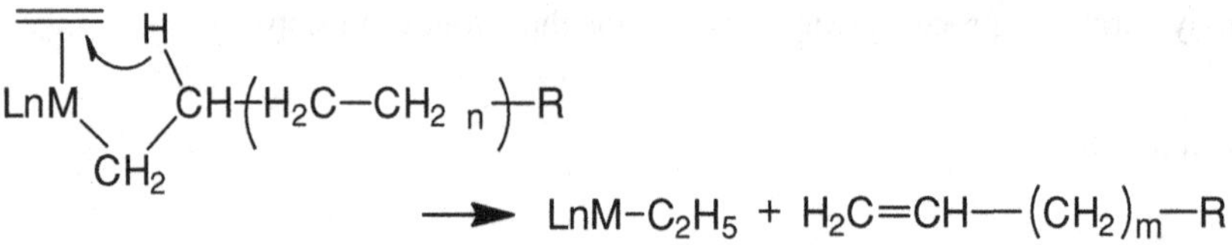

$$\longrightarrow \quad LnM-C_2H_5 \ + \ H_2C{=}CH—\left(CH_2\right)_m—R$$

Since, by construction, the polymeric chain has β hydrogens, the crucial point is to block the β-H elimination. We have seen earlier that the best electronic configuration for this purpose is d^0, hence the use of Ti(+4) and Zr(+4) catalysts. Higher numbers of d electrons favor oligomerization, or even, dimerization. In practice, some of the most useful catalysts derive from Cr(+4) and Zr(+4):

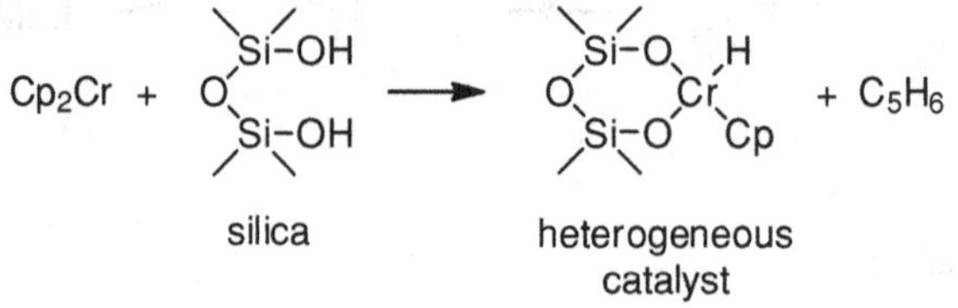

silica   heterogeneous
                                     catalyst

$$Cp_2ZrCl_2 \ + \ [MeAlO]_n$$

$$Cp_2ZrMe_2 \ + \ B(C_6F_5)_3 \ \xrightarrow{\text{pentane}} \ [Cp_2ZrMe] \ [MeB(C_6F_5)_3]$$

The replacement of ethylene by propene (also called propylene in the industry) introduces a problem of stereochemistry. Polypropylene exists in three varieties isotactic, syndiotactic and atactic, all with different mechanical properties:

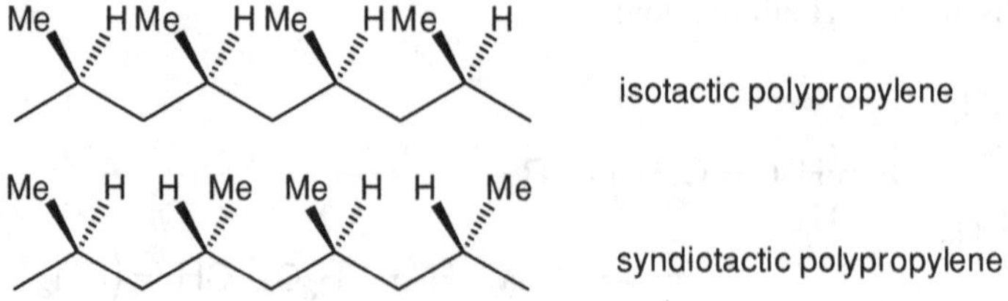

isotactic polypropylene

syndiotactic polypropylene

The atactic polypropylene has no defined stereochemistry and has a limited usefulness because its physical properties are poorly reproducible. It is possible to obtain either isotactic or syndiotactic polypropylene by choosing the appropriate zirconium catalysts:

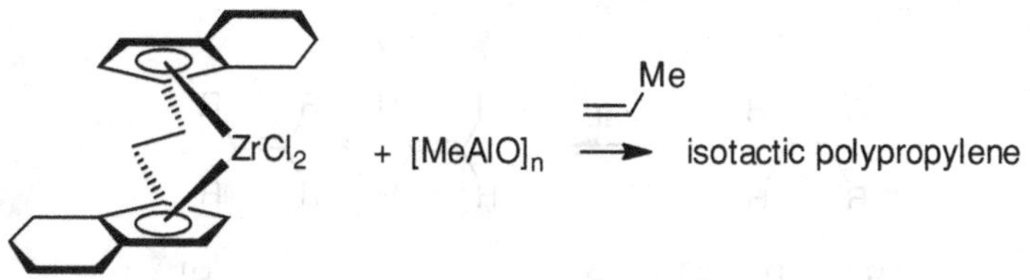

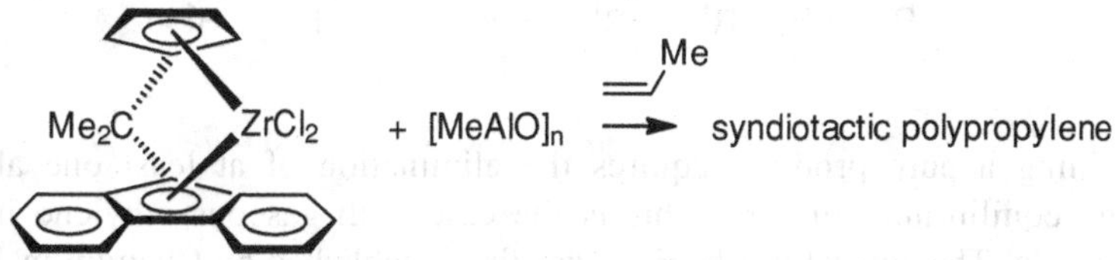

The bridges between the two η^5 ligands serve to block their rotation. In the first case, if we look to the zirconium center from the open side where the polymerization takes place, the propagation step takes place as shown:

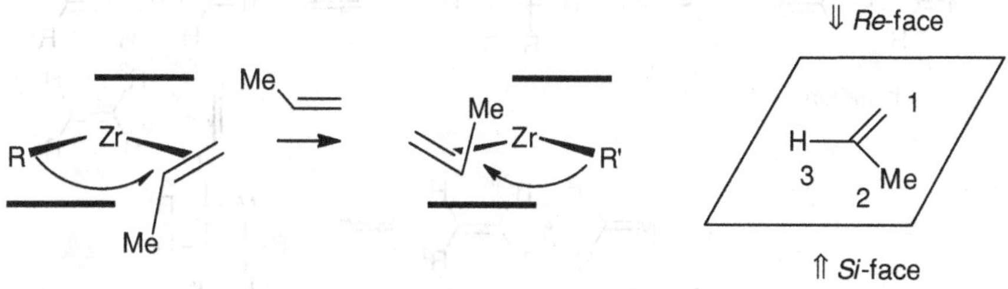

As can be seen, the polymeric chain always attacks the olefin on the same face (in this case, the *Si*-face), leading to the formation of the isotactic polymer. The situation is different in the second case:

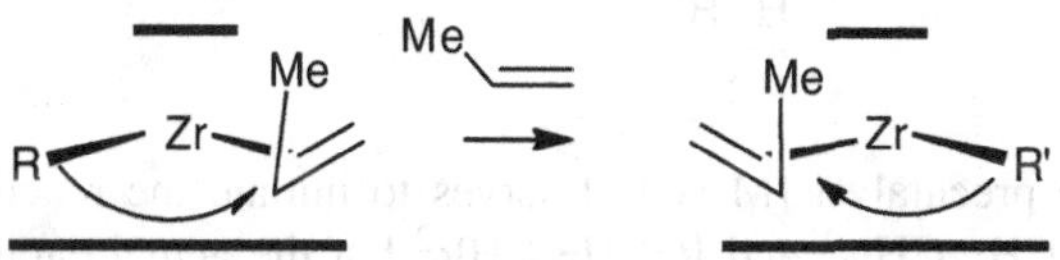

Here, the chain attacks on each face of the olefin alternatively, first the *Re*-face then the *Si*-face, leading to the formation of the syndiotactic polymer.

3.6 Alkene Metathesis [4]

If we consider alkenes as combinations of carbenes, then, the metathesis of alkenes is the statistical redistribution of the carbene fragments in the presence of a catalyst. The two following examples are representative:

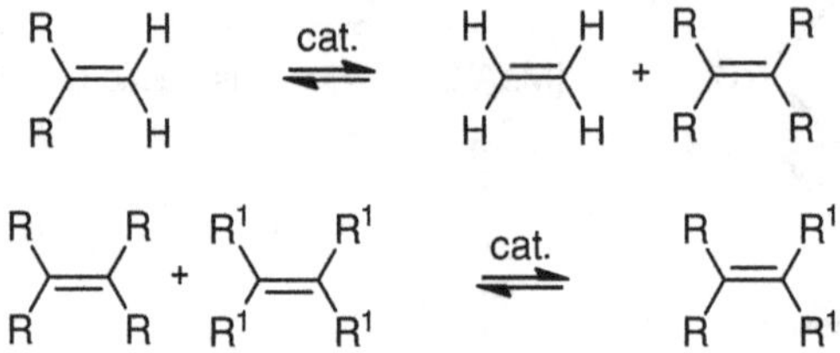

Obtaining a pure product requires the elimination of at least one alkene from the equilibrium mixture. This is the case with gaseous ethylene in the first example. The correct mechanism was first established by Chauvin in 1971. It involves a carbene complex as the catalyst and a metallacyclobutane as the intermediate:

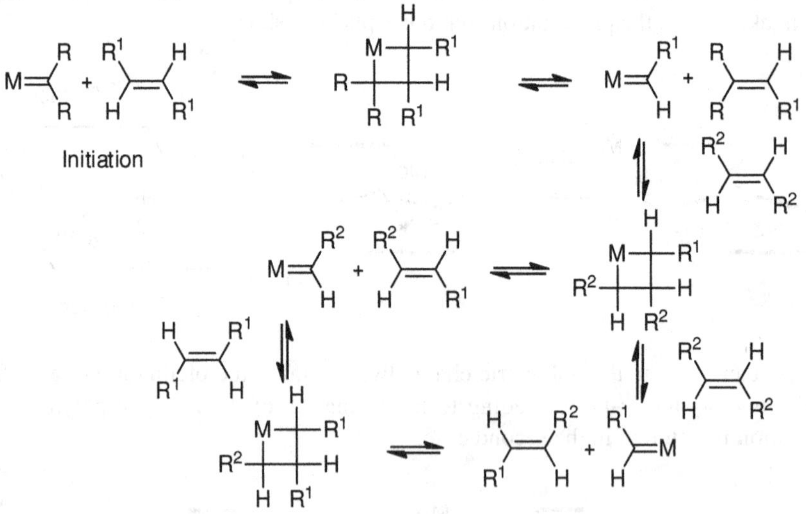

Initiation

The carbene precatalyst $[M=CR_2]$ serves to initiate the reaction between the two alkenes $R^1CH=CHR^1$ and $R^2CH=CHR^2$ but the actual catalysts that propagate the reaction are, alternatively, $[M=CHR^1]$ and $[M=CHR^2]$. Because they are both robust and efficient, the ruthenium-carbene catalysts of Grubbs are the most widely used today.

Cy = cyclohexyl, Mes = mesityl

In the Grubbs I catalyst, one phosphine is labile creating a vacant site for the coordination of the alkene which, then, forms the metallacyclobutane. The other highly basic and bulky phosphine serves to stabilize the 14e species. The role of the basic phosphine is taken by a *N*-heterocyclic carbene in the second case and by the even more basic saturated carbene in the third case.

The two most developed applications of the alkene metathesis are the ring closing metathesis (RCM) and the ring opening metathesis polymerization (ROMP).

This RCM is now the method of choice for the synthesis of macrocycles. It appears to perform well independently of the size of the macrocycle contrary to the classical methods.

The ROMP can serve to polymerize alkenes that cannot be polymerized using the Ziegler catalysts such as norbornene. The best synthesis of polyacetylene, whose use as an organic electroconducting material is well known, relies on the ROMP of cyclooctatetraene.

In the SHOP process (Shell Higher Olefin Process), a mixture of light (C_4 to C_9) and heavy (C_{15} to C_{40}) alkenes is submitted to a metathetic redistribution to give the more useful C_{10} to C_{14} alkenes used for the synthesis of detergents.

The metathesis of alkynes is also possible. In this case, the catalyst is a metal carbyne complex or a metal–metal triple bond (see below) and the intermediate a metallacyclobutadiene.

A specific deactivation pathway is available as shown:

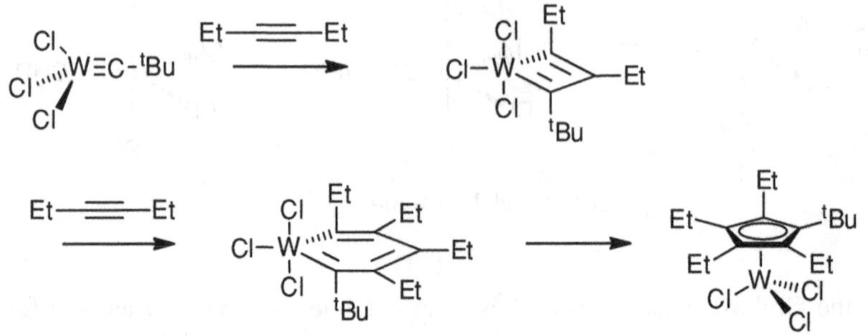

The aromaticity of the final η^5-cyclopentadienyl complex is the driving force of this deactivation. The metathesis of alkynes is now used in organic synthesis to build acetylenic macrocycles and polymers.

3.7 Palladium in Homogeneous Catalysis

Today, palladium is everywhere in organic synthesis as a result of its extraordinary ability to catalyze the formation of C–C bonds. But, in fact, the story started once again by an industrial process, the Wacker synthesis of acetaldehyde by oxidation of ethylene. The production of acetaldehyde by this process started in 1960. The stoichiometric reaction is shown here:

$$[PdCl_4]^{2-} + H_2C{=}CH_2 + H_2O \rightarrow CH_3CHO + Pd + 2\,HCl + 2\,Cl^-$$

Since palladium is reduced to Pd(0), it must be reoxidized to Pd(II) to transform this stoichiometric reaction into a catalytic process. This is achieved by using air and copper(I) chloride. The actual oxidation agent is copper(II) chloride which is reduced into copper(I) chloride. Finally, only oxygen is consumed. The mechanism of this reaction is quite complex. The key steps are shown here:

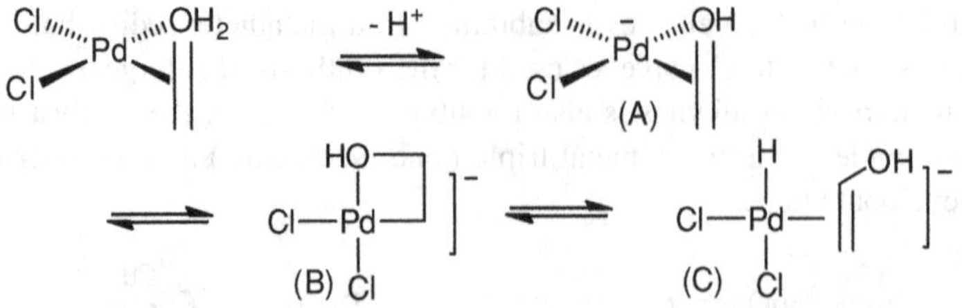

The transformation of complex (A) into complex (B) corresponds to a nucleophilic attack of OH^- onto ethylene. The transformation of complex (B) into complex (C) implies a decoordination of oxygen, followed by a β-H abstraction by Pd. After a series of additional steps, complex (C) collapses to give acetaldehyde, Pd, HCl, and Cl⁻. This chemistry can be transposed with other alkenes to give ketones:

$$R\diagdown\diagup \quad \xrightarrow[\text{DMF, H}_2\text{O}]{\text{PdCl}_2,\ \text{CuCl, O}_2} \quad R\overset{\text{O}}{\diagdown}\text{CH}_3$$

Water can be replaced by alcohols as shown in this intramolecular version:

$$R\diagup\diagdown\text{OH}\diagup\text{Me} \xrightarrow{\text{PdCl}_2,\ \text{CuCl}_2} \left[R\diagdown\!\overset{}{\underset{\text{Pd}}{\bigcirc}}\!\diagup\text{Me} \right] \xrightarrow{\text{CO, MeOH}} R\diagdown\!\overset{}{\bigcirc}\!\diagup\overset{\text{Me}}{\underset{\text{CO}_2\text{Me}}{}}$$

Alcohols can be replaced by amines but the problem is that amines coordinate strongly to palladium and tend to deactivate the catalyst. This has been finally solved recently as shown:

$$\bigcirc\!\!-\!\!\diagup + \text{PhNH}_2 \xrightarrow[\text{7 h, 100}^\circ\text{C}]{\text{Pd(OCOCF}_3)_2 + \text{L} + \text{TfOH}} \bigcirc\!\!-\!\!\overset{\text{NHPh}}{\diagup}\quad (99\%)$$

$$\text{L} = \begin{array}{c}\bigcirc\!\!-\!\!\text{PPh}_2\\ \text{Fe}\\ \bigcirc\!\!-\!\!\text{PPh}_2\end{array} \qquad \text{TfOH} = \text{CF}_3\text{SO}_3\text{H}$$

All of these catalytic reactions involve Pd(II). But, in fact, most of the palladium catalysts are based on Pd(0), typically the 14e species [PdL₂] where L is a variety of phosphines. This type of catalyst intervenes in the Heck, Stille, and Suzuki cross-coupling reactions. The Heck reaction [5], as initially described, creates a C–C bond between two sp² carbons:

$$\diagdown\!\!=\!\!\overset{\text{Br}}{} + =\!\!\diagup^{\text{CO}_2\text{Me}} + \text{Et}_3\text{N} \xrightarrow[100\ ^\circ\text{C}]{\text{[Pd(PPh}_3)_2]} \diagdown\!\!=\!\!\diagup\!\!=\!\!\diagdown^{\text{CO}_2\text{Me}} + [\text{Et}_3\text{NH}^+\text{Br}^-]$$

The mechanism involves the activation of the C–Br bond by oxidative addition:

$$\text{Ar-X} + \underset{14e}{\text{PdL}_2} \longrightarrow \underset{16e}{\overset{\text{Ar}_{\prime\prime\prime\prime}}{\underset{\text{L}}{\diagdown}}\overset{}{\underset{}{\text{Pd}}}\overset{\prime\prime\prime\prime\text{L}}{\underset{\text{X}}{\diagup}} \quad trans \text{ addition (S}_\text{N}2)$$

$$\underset{}{\overset{}{\underset{\text{Z}}{=\!\!\diagup}}} \longrightarrow \underset{18e}{\overset{\text{Ar}_{\prime\prime\prime\prime}\text{Pd}^{\prime\prime\prime\prime}\text{L}}{\underset{\text{L}\diagdown\!\!\text{X}}{\diagup\ \ \diagdown\text{Z}}}} \xrightarrow{cis \text{ migration}} \underset{\text{Ar}}{\overset{\text{X}\diagdown\ \diagup\text{L}}{\underset{\text{L}}{\overset{\text{Pd}}{\diagup}}\diagdown\text{Z}}}$$

The ionization of the Pd–X bond (X = OTf) is highly favorable because it avoids the formation of an 18e complex. Similarly, very bulky phosphines that easily leave the coordination sphere to relieve steric crowding have a positive effect. The end of the catalytic cycle involves a β-H elimination:

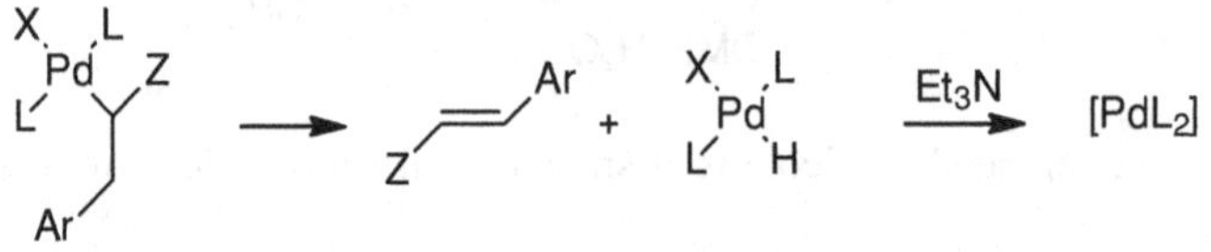

The reaction is, in practice, restricted to the coupling of sp^2 carbons in order to avoid undesired β-H eliminations during the catalytic cycle. The problems that are currently under investigation concern the coupling of the cheap yet unreactive aryl chlorides and the asymmetric version of the Heck reaction. The use of the very bulky and very basic tris-*tert*-butylphosphine has brought a partial solution to the first problem:

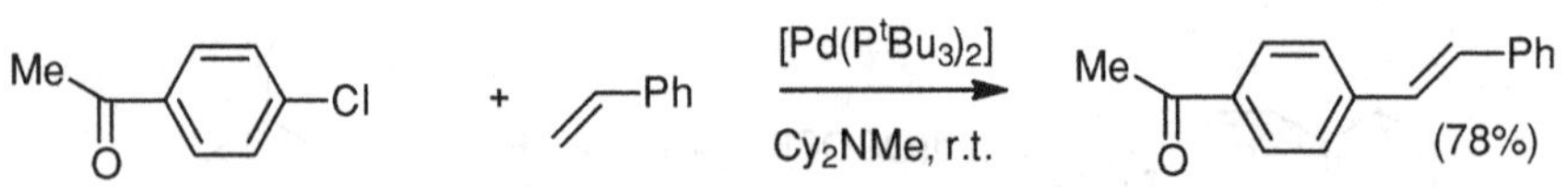

The very bulky and very nucleophilic *N*-heterocyclic carbenes are also a possibility. The test reaction for the asymmetric version is shown here:

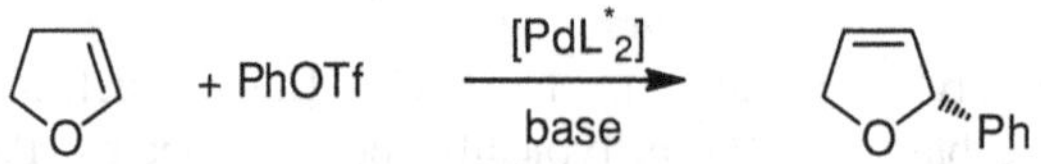

In the final β-H elimination, the palladium is on the β carbon on the back of the ring as is the phenyl substituent. It can only interact with the appropriate hydrogen on the β' position.

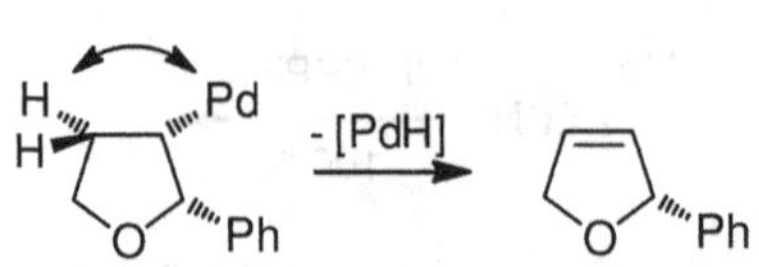

E.e.'s as high as 96 % have been obtained with BINAP or phosphinooxazolines as the optically pure ligands.

The Stille reaction relies on the coupling of a mild carbon nucleophile with the aryl bromide as shown:

$$Ar-X + Bu_3Sn-R \xrightarrow{[PdL_2]} Ar-R$$

Since tin bears four carbon substituents, the reaction will be useful only if there is a well-defined transfer order from tin to palladium. This is indeed the case:

$$RC \equiv C- \; > \; RCH{=}CH- \; > \; Ph \; > \; PhCH_2 \; > \; Me \; > \; Bu$$

After the initial activation of Ar–X by Pd(0), the tin derivative reacts with the Pd–X bond:

This coupling reaction can be modified to obtain ketones and aldehydes:

$$RC(O){-}X + Bu_3Sn{-}R' \xrightarrow{[PdL_2]} RC(O)R' + Bu_3Sn{-}X$$

$$RX + CO + Bu_3SnH \xrightarrow{[PdL_2]} RC(O)H + Bu_3Sn - X$$

An interesting application of the Stille cross-coupling reaction is shown here. The bromide does not derive from an sp^2 carbon as usual but is devoid of β-hydrogens.

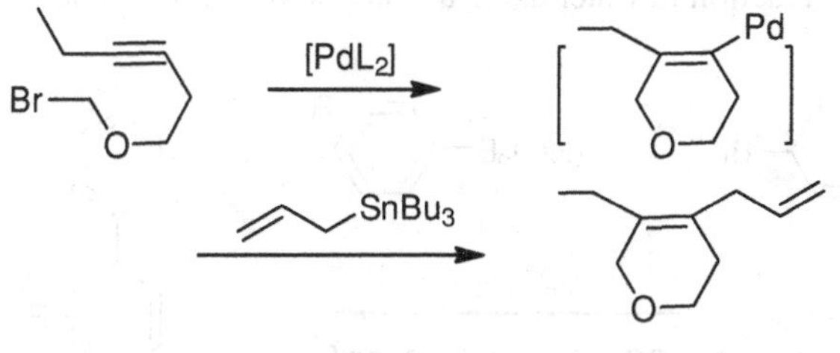

In the Suzuki cross-coupling reaction [6], the tin nucleophile is replaced by an anionic tetracoordinate boron nucleophile. The overall scheme is shown here:

$$Ar{-}X + RB(OH)_2 \xrightarrow[\text{[R'O}^-]]{[PdL_2]} Ar{-}R$$

The boronic acid itself is not nucleophilic. The oxygen nucleophile $R'O^-$ transforms it into a tetracoordinate borate which is negatively charged at boron. The second role of this oxygen nucleophile is to convert the Pd–X into a Pd–OR' bond. This Pd–O bond is weaker and more reactive than the Pd–X bond because palladium is more halophilic than oxophilic. The actual reaction is shown here:

The Suzuki reaction has proved to be extraordinarily versatile and is heavily used today to build C–C bonds. A highly congested and highly basic ligand has been developed by Buchwald for the coupling of aryl chlorides:

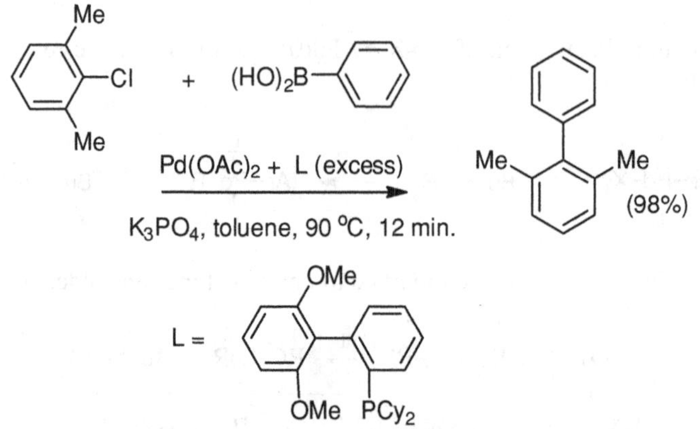

The ligand was conceived in order to have a maximum of electronic density and steric hindrance at phosphorus. The OMe groups also prevent a possible orthometallation at the expense of the *ortho* C–H bonds. The likely catalyst in this case is [LPd] in which Pd is loosely chelated between P and O. It is also possible to perform the Suzuki reaction in water using a water-soluble phosphine as the ligand.

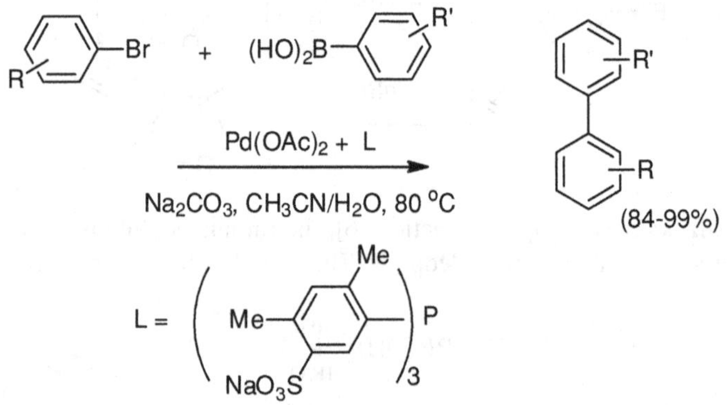

Among the numerous other applications of palladium in organic synthesis, one particularly stands out. It relies on the use of the easily accessible η^3-allyl-palladium species. Their synthesis is shown here:

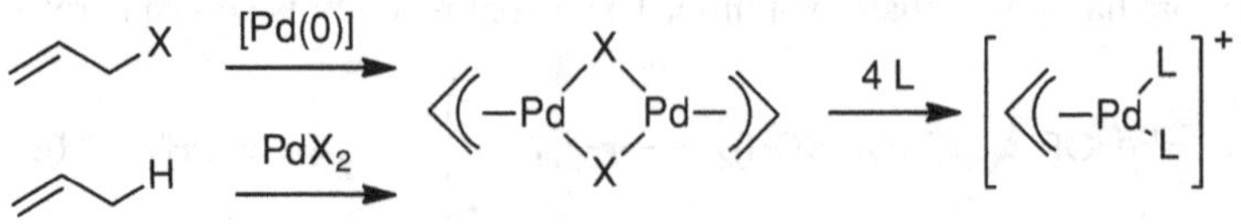

The final cationic complex easily reacts with nucleophiles at the terminal positions and on the *exo* side to give the organic products while releasing the [PdL$_2$] catalyst.

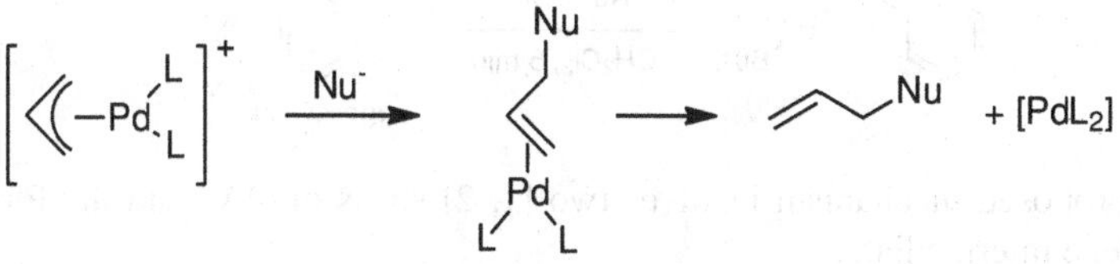

This scheme has been extensively used in organic synthesis.

3.8 Gold in Homogeneous Catalysis [7]

For a long time, gold was believed to be useless in homogeneous catalysis. Then, around 2000, it was shown that gold could have interesting catalytic applications, for example in the diboration of alkenes. It was a turning point and, now, gold is one of the most studied metals in catalysis. It is, of course, necessary to wait in order to discriminate between the really useful applications and the background "noise".

Gold chemistry has three peculiarities which are significant for catalysis: (1) The [LAu]$^+$ cation (L = PR$_3$) is isolobal with the proton (same frontier orbitals); (2) As a result of the relativistic speed of its external electrons, a gold atom creates an attractive interaction with another gold atom close to it. This is the so-called aurophilic interaction; (3) Gold is the most electronegative of all the metals and is easily reduced to Au$^-$.

For its catalytic applications, gold is generally coordinated with phosphines and stabilized singlet carbenes (NHC). A typical synthesis of a gold-carbene complex is shown here:

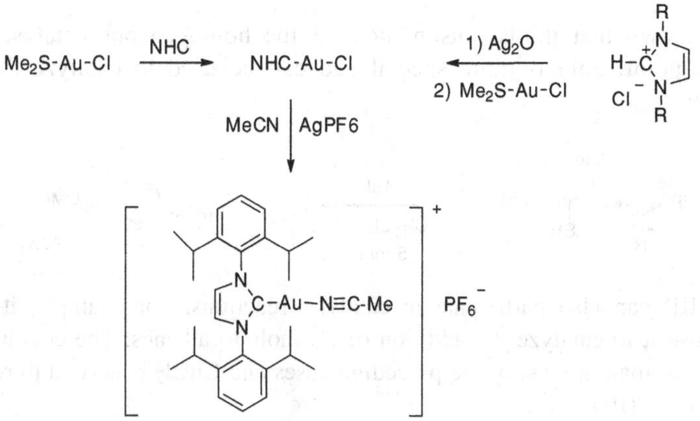

Gold is especially prone to interact with alkynes and has been shown to catalyze several of their transformations such as their cycloisomerisations.

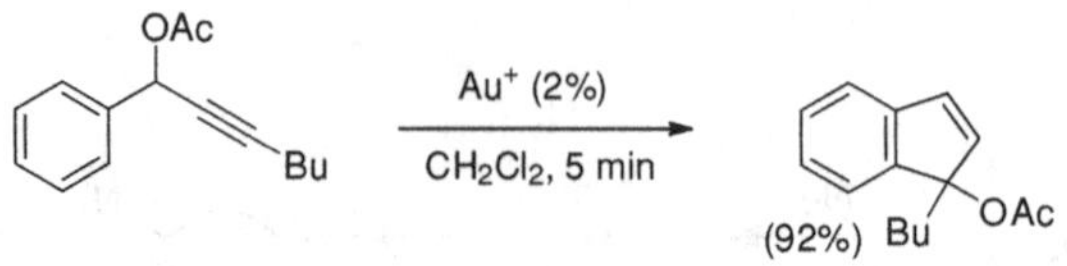

The proposed mechanism involves two [1, 2] shifts of OAc and the formation of an allene intermediate.

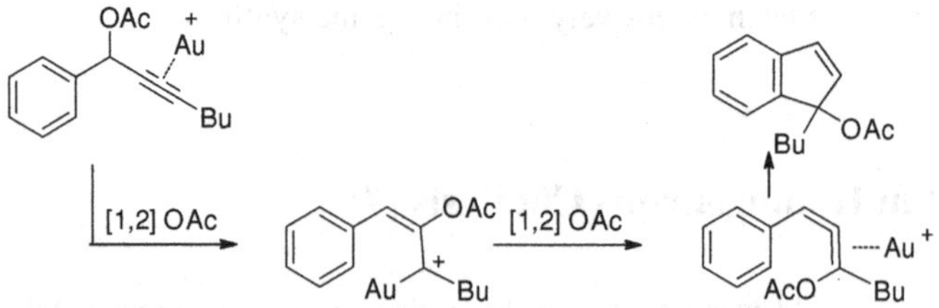

The intramolecular hydroamination of alkynes is also quite easy:

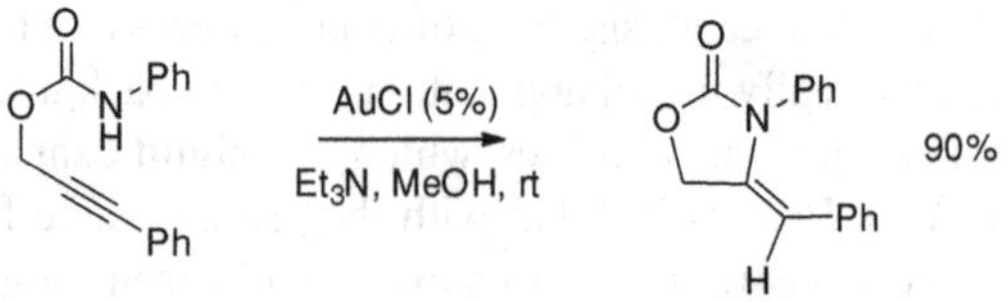

The intermolecular hydroamination is much more difficult and needs high temperature and microwave irradiation:

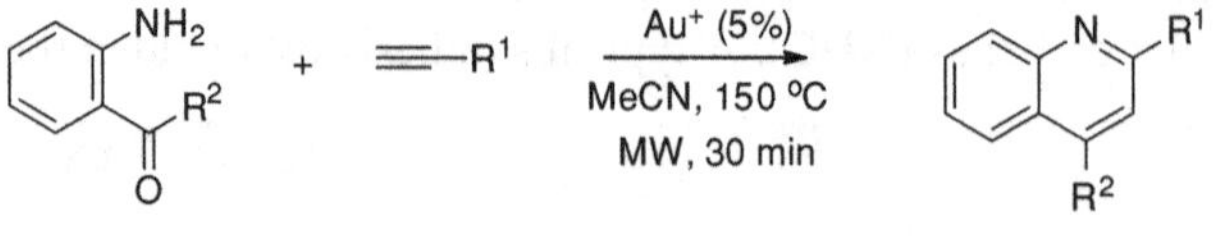

It is known that thiols poison most of the homogeneous catalysts. From this standpoint, gold is quite special and can be used to catalyze thiolation reactions:

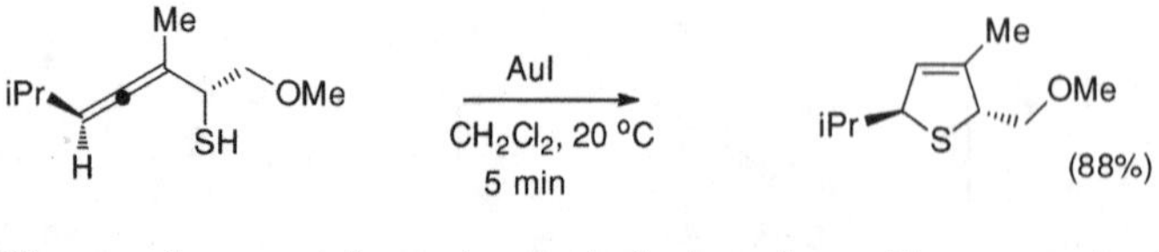

Gold(III) can also participate in catalytic reactions. For example, it is possible to use it to catalyze the addition of alcohols to alkenes. The conditions are more drastic than in most of the preceding cases and $CuCl_2$ is needed to reoxidize gold(I) to gold(III).

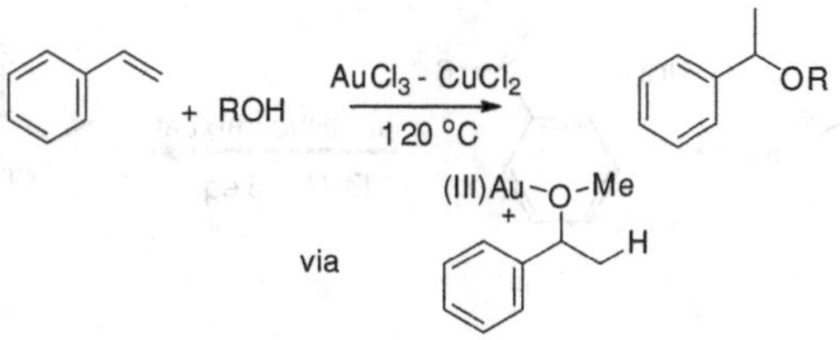

Gold(III) can also serve to activate C–H bonds, both in intra and intermolecular reactions.

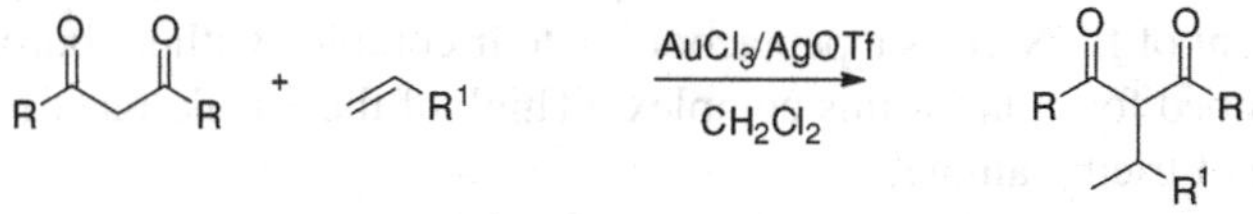

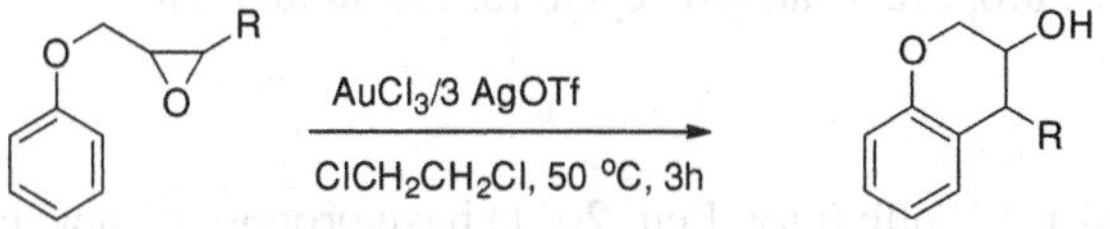

The example below combines the hydroamination of the aldehyde, the condensation with the alkyne C–H bond with elimination of water, and the cyclization promoted by the gold catalyst:

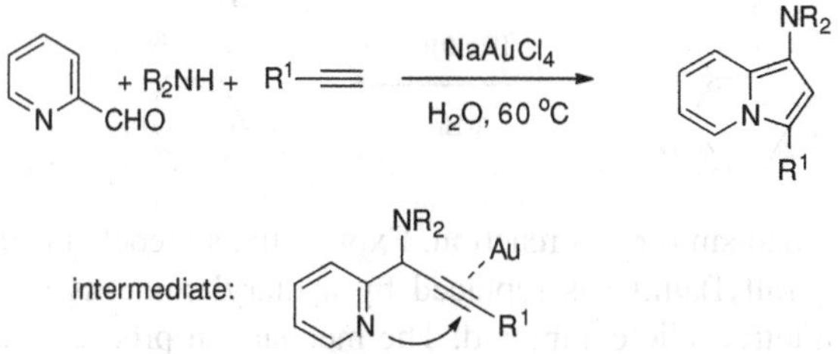

Gold(I) is also known to catalyze the hydrogenation of functional alkenes, as well as the reduction of ketones and nitro groups, but these transformations are less original. Numerous other catalytic uses of gold have been unveiled in the literature but they are clearly outside the scope of this introductory textbook.

3.9 Problems

III.1

(T. Nishimura et al. J. Am. Chem. Soc., 2007)
The paper describes the iridium-catalyzed annulation of 1,3-dienes with ortho-carbonylated phenylboronic acids. The overall scheme is given below.

(1) Propose a developed formula for the iridium catalyst (cod = 1,5-cyclooctadi-
    ene). What is the oxidation state of Ir and the number of electrons on Ir?
(2) The stoichiometric reaction of the Ir complex with one equivalent of ortho-
    formylphenylboronic acid and one equivalent of PPh$_3$ in the presence of one
    equivalent of Et$_3$N gives a pentacoordinate Ir complex without boron. Propose
    a developed formula for this complex. (Think of the Suzuki reaction). What is
    the role of triethylamine?
(3) On this basis, propose a catalytic cycle for the annulation.

III.2

D.M Schultz and J.P. Wolfe (Org. Lett. 2011) have proposed a new method for the
synthesis of polycyclic nitrogen heterocycles based on aminopalladation and car-
bometallation cascade reactions:

Propose a mechanism for this reaction. Explain the stereochemistry.

When the aryl(allyl)amine is replaced by a diarylamine as shown, the same
reaction yields a tetracyclic compound. The mechanism proceeds initially via the
same kind of intermediate

Propose a mechanism for this reaction. Explain the stereochemistry.

III.3

A paper (Duric, Tzschucke, Org. Lett. 2011) has described the following synthesis
of bipyridines:

(1) What is the actual catalyst and why has tris(tertbutyl)phosphine been chosen as the ligand?

(2) Why has pyridine N-oxide been chosen as a co-reagent instead of pyridine itself?

(3) Propose a mechanism. If a para-CO_2Et substituent is present on the pyridine N-oxide, a tricyclic product is obtained. Propose a formula for this product and explain its formation.

III.4

The following transformation has been described (Uenishi Org. Lett. 2011):

(1) The presence of the double bond is necessary for the transformation to proceed. Why?

(2) Which is the most electron-rich oxygen?

(3) The catalytic intermediate is a chelate. Give its formula.

(4) On this basis, propose a mechanism.

III.5

The following catalytic transformation has been performed by Madsen (Organometallics 2011):

A ruthenium dihydride is the actual catalyst. How is it formed in the reaction medium? Once it is formed, it reacts with RCH_2OH to release hydrogen and give an aldehyde complex (A). What is the formula of this complex (A) and how is it formed? (A) reacts with a second molecule of RCH_2OH with the release of a second molecule of H_2 and forms (B). What is the formula of (B)? (B) gives the final product and regenerates the dihydride catalyst. What is the reaction involved?

III.6

A paper (Allegretti, Ferreira, Org. Lett. 2011) has described the following synthesis of furans:

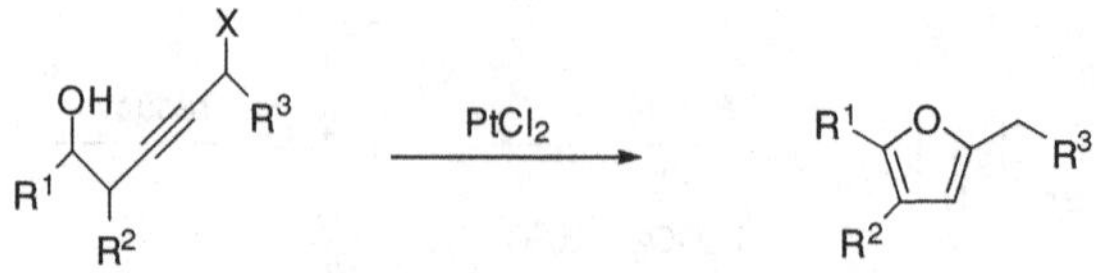

(1) The proposed mechanism first involves the complexation of platinum to give complex (A). What is the formula of (A)?

(2) (A) cyclizes to give a 5-membered ring (B). What is the formula of (B)? What is the type of reaction involved ?

(3) HX is eliminated and a platinum–carbene complex (C) is formed. What is the formula of (C)?

(4) A shift of hydrogen takes place and Pt is released. What is the formula of the resulting organic product (D)? Why does it evolve to give the final furan?

Throughout the reaction scheme, use [Pt] as a symbol for the catalytic center.

III.7

A published work describes the direct acylation of aryl C–H bonds by alcohols using $PdCl_2$ as a catalyst and tBuOOH as the stoichiometric oxidant. The overall reaction is shown below.

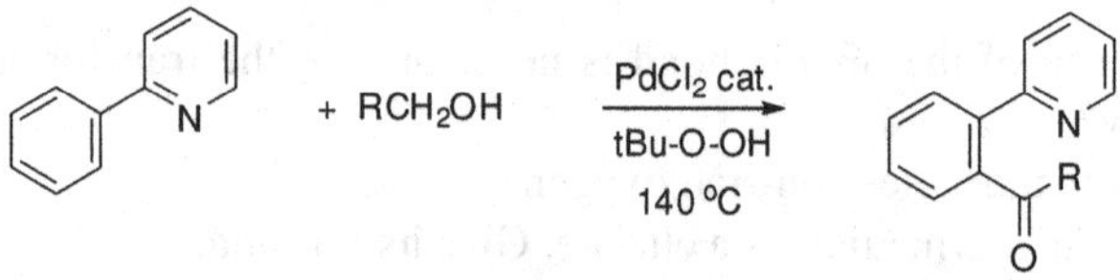

The reaction is run at 140 °C in chlorobenzene. At this temperature, the actual catalyst is unknown. We just know that it derives from Pd(II). The mechanism involves 4 steps. The oxidant serves both to oxidize the alcohol and, at one point, the Pd intermediate. Propose a mechanism. Explain the role of nitrogen. How are the palladium intermediate and the alcohol transformed by the oxidant? Each step must include the type of reaction and the oxidation state of the metal.

III.8

S.C. George et al. Org. Lett. 2011

In your opinion, is heptafulvene **1** aromatic, antiaromatic, or nonaromatic? What is the expected consequence for its reactivity?

In the presence of a Pd(II) precatalyst, **1** reacts with allyl chloride and allyl tributyltin to give **2**

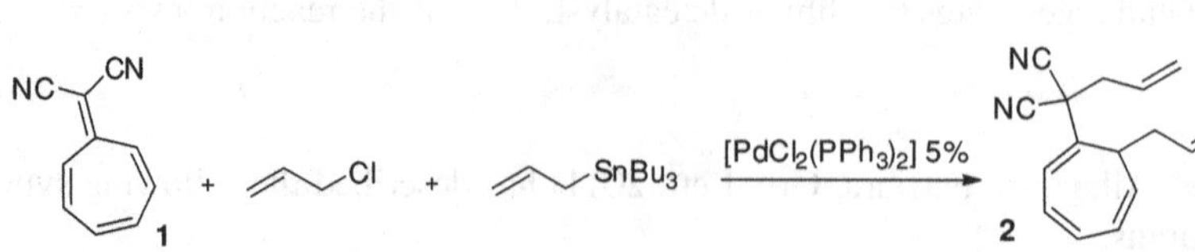

What is the actual catalyst?

Propose a catalytic scheme.

From the standpoint of organic chemistry, what type of addition reaction is this?

III.9

The following transformation has been described (Luo, Wu Org. Lett. 2011):

(1) The first step involves the formation of a chelate of Pd(II) (A). What is the formula of (A)?

(2) The second step involves the incorporation of CO in the presence of the base to give (B). What type of reaction is this? What is the role of the base? What is the formula of (B)?

(3) A second molecule of alkyne is coordinated to Pd. The base induces the formation of the benzofuran ring (right part of the molecule) to give (C). What is the formula of (C)? What is the type of reaction involved?

(4)   (C) evolves to give the final product. What is the type of reaction involved? Why is oxygen necessary to close the catalytic cycle?

References

1. Noyori R (2008) Asymmetric catalysis: science and opportunities (Nobel lecture). Angew Chem Int Ed 2002:41

2. Beller M, Cornils B, Frohning CD, Kohlpaintner CW (1995) Progress in hydroformylation and carbonylation. J Mol Catal A Chem 104:17

3. Bochmann M (1996) Cationic group 4 metallocene complexes and their role in polymerization catalysis: the chemistry of well-defined Ziegler catalysts. J Chem Soc Dalton Trans 255

4. Grubbs RH, Chang S (1998) Recent advances in olefin metathesis and its application in organic synthesis. Tetrahedron 54:4413

5. Beletskaya IP, Cheprakov AV (2000) The Heck reaction as a sharpening stone of palladium catalysis. Chem Rev 100:3009

6. Suzuki A (1999) Recent advances in the cross-coupling reactions of organoboron derivatives with organic electrophiles. J Organomet Chem 576:147

7. Hashmi A, Stephen K, Hutchings GJ (2006) Gold catalysis. Angew Chem Int Ed 45:7896

Solutions to the Problems

I.1

$ReH_9{}^{2-}$ 18e, OS +7, d^0

    $TaMe_5$ 10e, OS +5, d^0

    $[(Ph_3P)_3Ru\ (\mu\text{-}Cl)_3\ Ru(PPh_3)_3]^+$ The bridge counts for 9e from which the positive charge is deduced. Hence, each Ru gains 4e from the bridge. Overall: 18e, OS +2, d^6.

I.2

$MeReO_3$ 14e, OS +7, d^0

    $CpMn(CO)_3$ 18e, OS +1, d^6

    $[Re_2Cl_8]^{2-}$ The $ReCl_4{}^-$ unit has 12e. A 16 e configuration with a Re–Re quadruple bond is likely.

I.3

In M–Cl, chlorine has still three available lone pairs. $Re(CO)_3Cl$ has 14e and needs to use two additional lone pairs. Hence, chlorine must act as a μ_3 ligand and the compound is a tetramer:

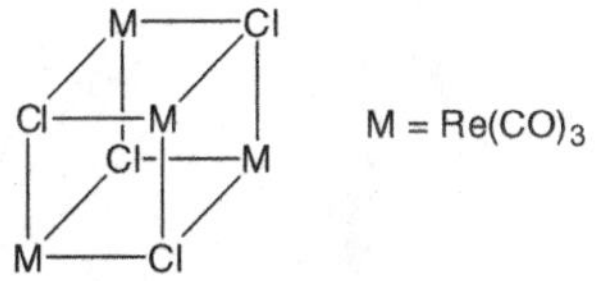

M = Re(CO)₃

F. Mathey, *Transition Metal Organometallic Chemistry*,
SpringerBriefs in Molecular Science, DOI: 10.1007/978-981-4451-09-3,
© The Author(s) 2013

I.4

Ph_2P has three available electrons for complexation. The possible complexes are:

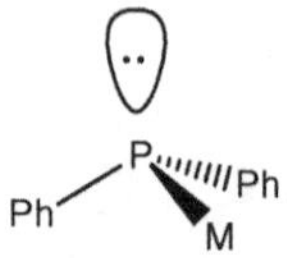

1e terminal ligand
pyramidal, one lone pair

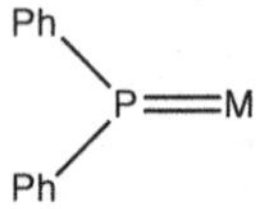

3e terminals igand,
planar, P=M double bond
no lone pair

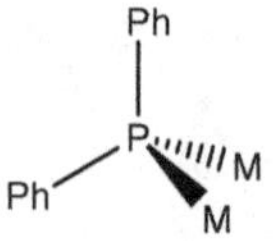

3e, bridging ligand,
tetrahedral, no lone pair

PhP has four available electrons for complexation. The possible complexes are:

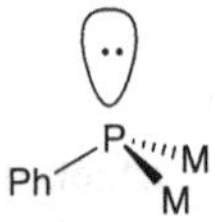

2e bridging ligand,
pyramidal, one lone pair

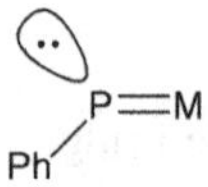

2e terminal ligand, bent,
P=M double bond
one lone pair

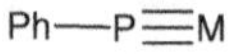

4e terminal ligand, linear
P≡M triple bond
no lone pair

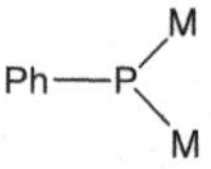

4e bridging ligand, planar,
no lone pair

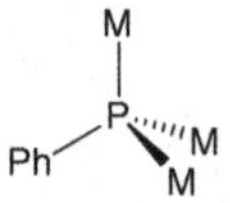

4e bridging ligand,
tetrahedral, no lone pair

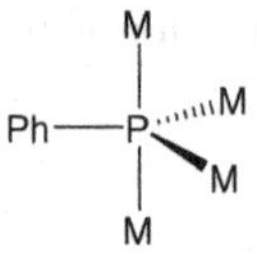

4e bridging ligand,
TBP, no lone pair

I.5

$(OC)_3Co(NO)$ has an 18e configuration with NO acting as a 3e ligand. The Co–NO unit is linear. The formal oxidation state of cobalt is -1 because NO is considered as NO^+.

I.6

Nickelocene is a 20e complex. It fluctuates between $(\eta^5-Cp)_2Ni$ and (η^5-Cp) Ni (η^1-Cp) (16e). The 16e complex can add L to give (η^5-Cp) Ni $(\eta^1-Cp)L$. The σ bond Ni- (η^1-Cp) reacts with IMe to give Me–Cp + CpNi(I)L.

I.7

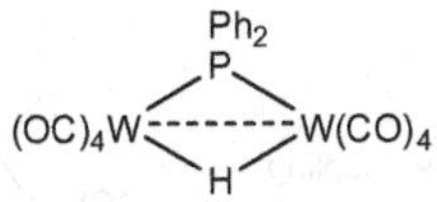

$$
\begin{array}{c}
Ph_2 \\
P \\
(OC)_4W \text{------} W(CO)_4 \\
H
\end{array}
$$

The bridges count for 4e. Without the metal–metal bond, the tungsten atom has 16e. The OS is +1. A W–W double bond is possible. The reaction path is probably:

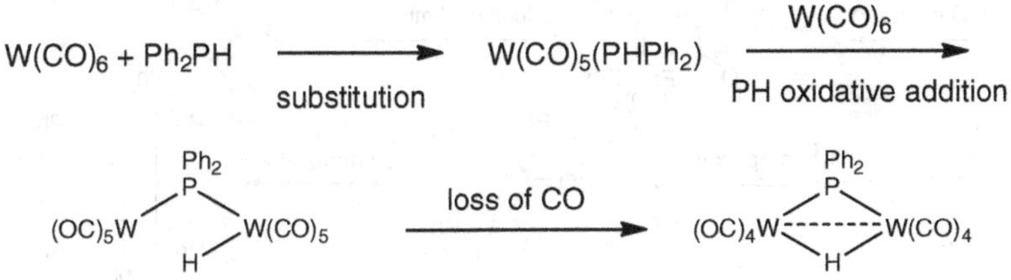

$$W(CO)_6 + Ph_2PH \xrightarrow{\text{substitution}} W(CO)_5(PHPh_2) \xrightarrow[\text{PH oxidative addition}]{W(CO)_6}$$

$$
\begin{array}{c}
Ph_2 \\
P \\
(OC)_5W \qquad W(CO)_5 \\
H
\end{array}
\xrightarrow{\text{loss of CO}}
\begin{array}{c}
Ph_2 \\
P \\
(OC)_4W \text{------} W(CO)_4 \\
H
\end{array}
$$

I.8

If NO is a 1e ligand, $[Fe(CN)_5NO]^{2-}$ is a 16e complex and Fe has the +4 oxidation. This is not likely. If NO is a 3e ligand, Fe has 18e and the OS is +2. This is the correct formulation. Fe–NO is linear and the salt diamagnetic.

I.9

The two elementary steps are insertion and β-H elimination:

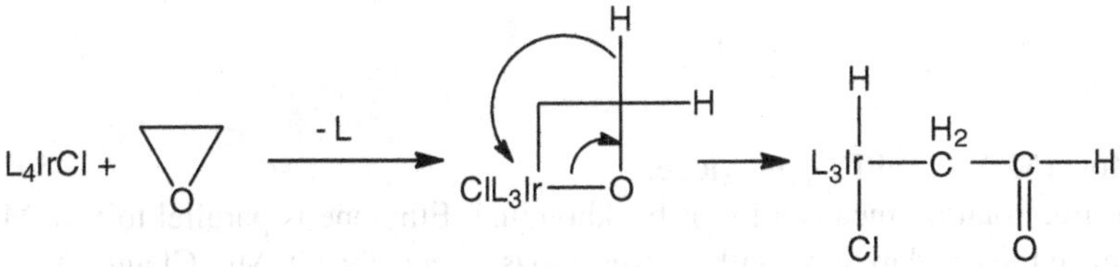

The oxidation state of Ir is +1 before and +3 after the reaction.

I.10

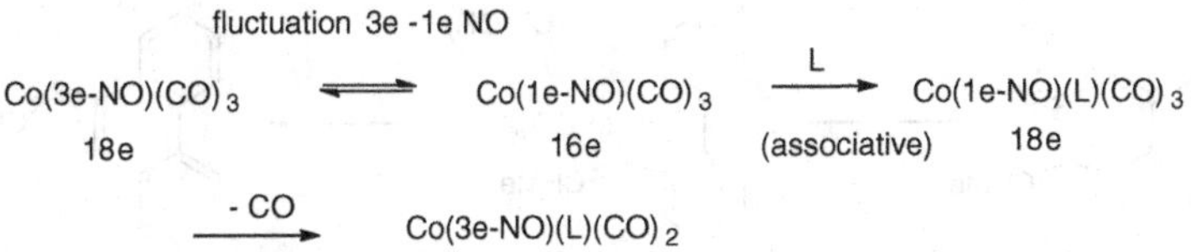

fluctuation 3e -1e NO

$$Co(3e\text{-}NO)(CO)_3 \rightleftharpoons Co(1e\text{-}NO)(CO)_3 \xrightarrow[\text{(associative)}]{L} Co(1e\text{-}NO)(L)(CO)_3$$

18e   16e   18e

$$\xrightarrow{-CO} Co(3e\text{-}NO)(L)(CO)_2$$

I.11

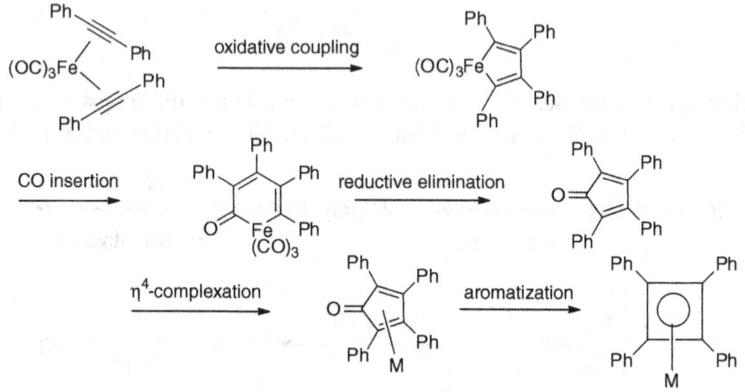

I.12

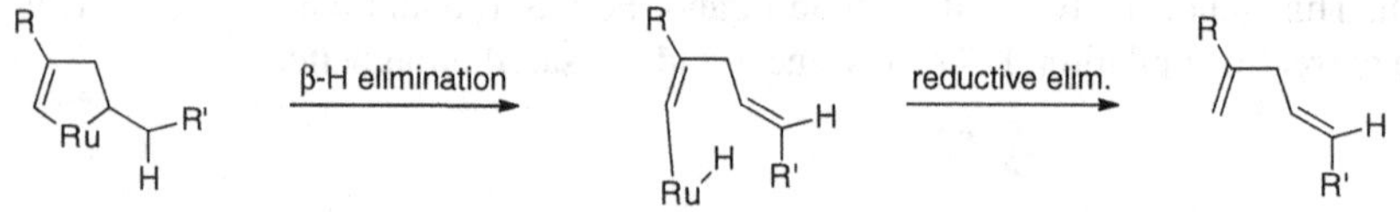

idem from:

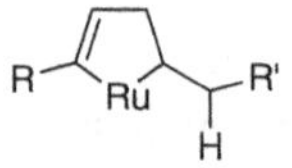

II.1

$Cp_2MoCl(Me) + AgPF_6 +$ ethylene.

No free rotation means a lot of backbonding. Ethylene is parallel to Mo–Me to maximize the overlap of π^* with d_{yz} (the z axis bisects the Cl–Mo–Cl angle).

The two CH_2 are not equivalent in ^1H and ^{13}C NMR.

II.2

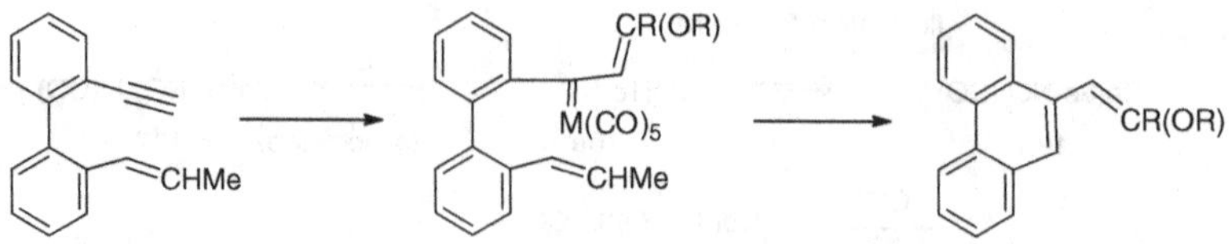

II.3

A $\left[(OC)_5Cr{-}\overset{\displaystyle C{\equiv}C{-}R^1}{\underset{\displaystyle R}{\overset{\displaystyle |}{C}}}{-}OMe \right]^{-}$ Li$^+$ B $(OC)_5Cr{=}\overset{\displaystyle C{\equiv}C{-}R^1}{\underset{\displaystyle R}{\overset{\displaystyle |}{C}}}$

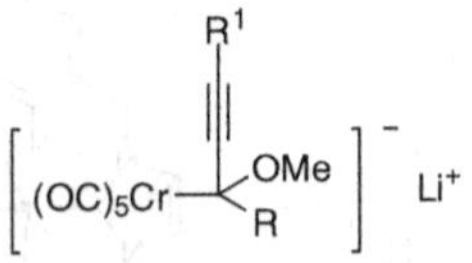

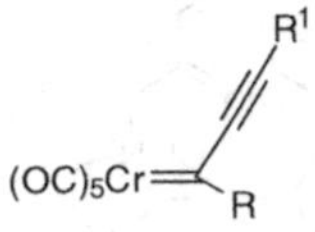

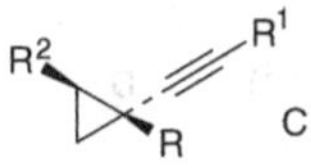

C (two stereochemistries)

II.4

B $\underset{\displaystyle OC}{\overset{\displaystyle OC\cdots}{}}\overset{\displaystyle Cp}{\underset{\displaystyle}{\overset{\displaystyle |\;+}{Mn}}}{=}\overset{\displaystyle}{\underset{\displaystyle R}{C}}{-}PR_2H$ C $\underset{\displaystyle OC}{\overset{\displaystyle OC\cdots}{}}\overset{\displaystyle Cp}{\underset{\displaystyle}{\overset{\displaystyle |}{Mn}}}{=}\overset{\displaystyle}{\underset{\displaystyle R}{C}}{-}PR_2$

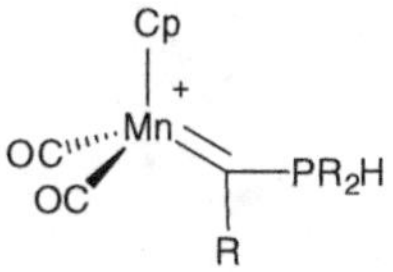

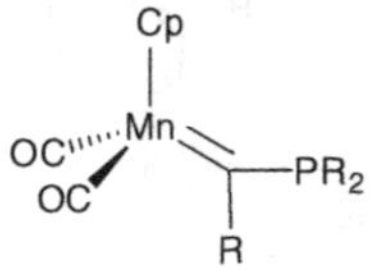

C → D similar to the conversion of Fischer carbenes into η^2-ketene complexes

II.5

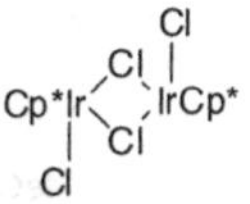

Cp*Ir$\overset{\displaystyle Cl}{\underset{\displaystyle Cl}{}}$Cl IrCp*   18e   Cp*Ir$\overset{\displaystyle O}{\underset{\displaystyle AcO}{}}$CMe

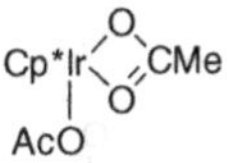

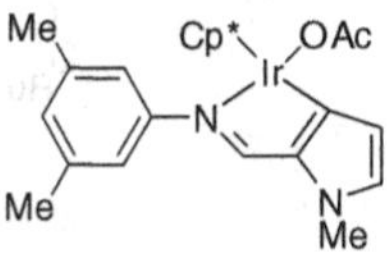

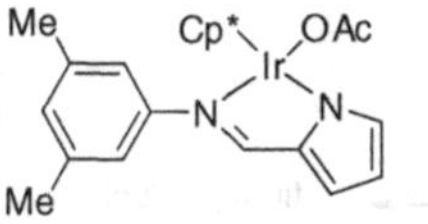

Me, Cp*, OAc ... A

Me, Cp*, OAc ... B

II.6

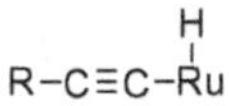

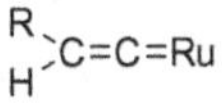

$R{-}C{\equiv}C{-}\overset{\displaystyle H}{\underset{\displaystyle}{Ru}}$ $\overset{\displaystyle R}{\underset{\displaystyle H}{}}C{=}C{=}Ru$ (migration of H from Ru to C)

The attack of the first complex by the carboxylate gives:

$R^1{-}\overset{\displaystyle O}{\overset{\displaystyle ||}{C}}{-}O{-}\overset{\displaystyle}{\underset{\displaystyle}{C}}{=}CH_2 \; R$

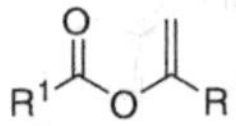

The attack of the carbene complex by the carboxylate gives:

$R^1{-}\overset{\displaystyle O}{\overset{\displaystyle ||}{C}}{-}O{-}CH{=}CH{-}R$

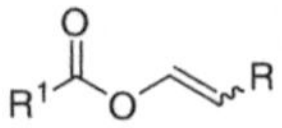

II.7

Cl
Cl
Ir
N
R
C (16e)
H
H
H2
D
PMe3
E
F
H fluxional in D

II.8

A
Ru
H^+
B
MeO⁻
C
final product

II.9

Ph
ᵗBu
Fe
(CO)3
BuLi
CO
(OC)3Fe
Bu
metathesis
- BuCO2⁻
Fe(CO)3
CO

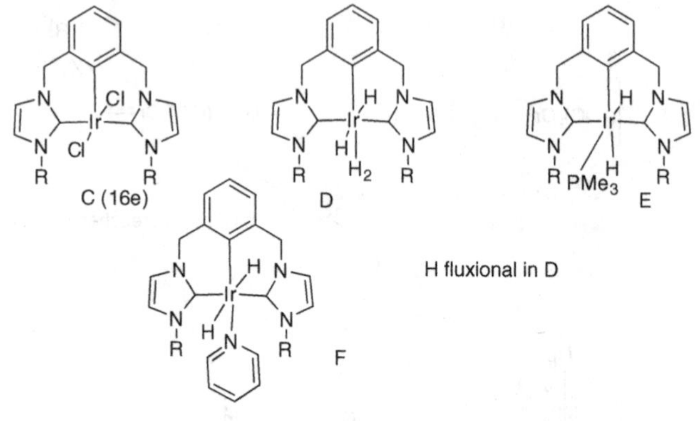

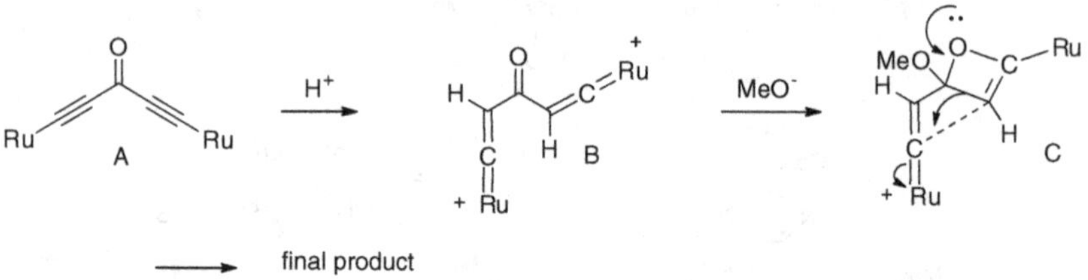

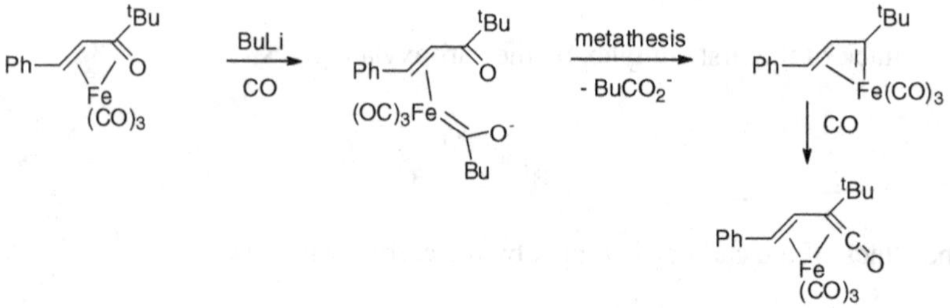

II.10

The active species is benzyne-zirconocene.

III.1

(1)

Ir: 16e, OS +2

(2)

(3)

Ir ≡ Ir(cod); the diene replaces PPh₃

III.2

III.3

The catalyst is probably [Pd(P^{t}Bu$_3$)$_2$] stabilized by the bulky phosphine.
The α-CH's of pyridine N-oxide are easily metallated.

the ester group facilitates the
double metallation

III.4

The double bond coordinates Pd.
  The most reactive oxygen is the epoxide oxygen.

III.5

III.6

(nucleophilic attack
on coordinated ligand)

(D) gives the more stable aromatic furan by [1,5] H migration

III.7

Coordination of nitrogen to Pd(II) and ortho-metallation:

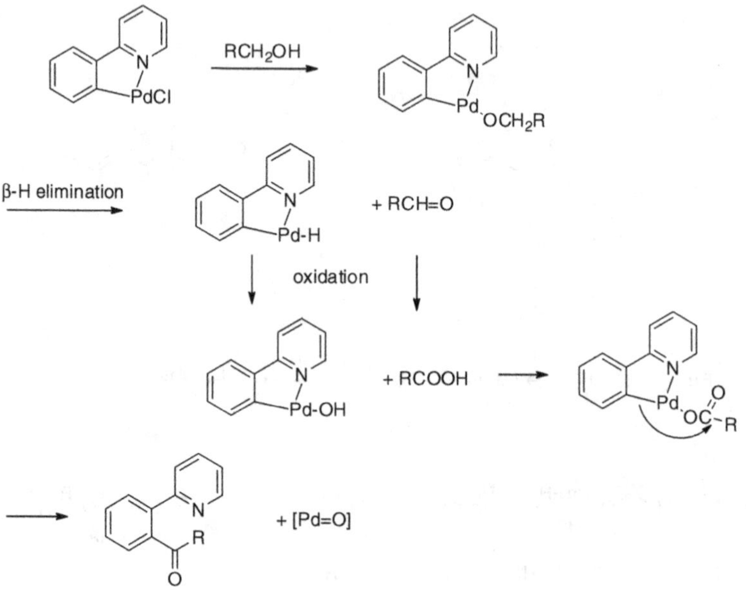

III.8

Taking into account the polarization of the exocyclic double bond, this heptaful-vene is somewhat aromatic.

The actual catalyst is a bis (η^3-allyl) palladium.

The reaction is a conjugate 1,8-addition.

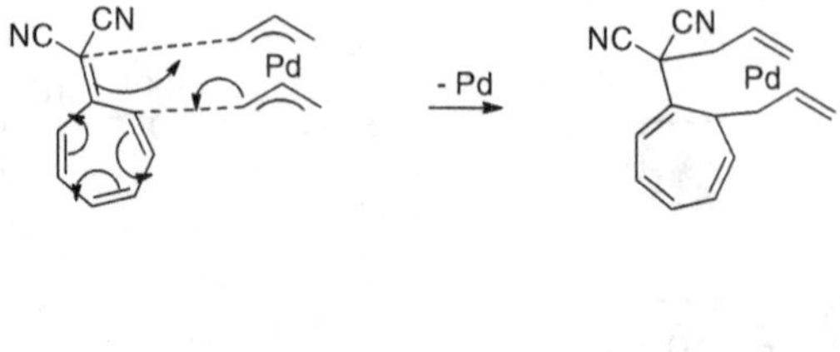

III.9

Oxygen is necessary to reoxidize Pd(0) to Pd(II) which is the actual catalyst.

Index